Physics Essays for Advanced Pupils

Dr Brian M. E. de Silva & Philip J. Lawson

DEDICATION

Brian dedicates his authorship of this text to the loving memory of his late parents G. R. Walter & Geraldine. N. C. de Silva

Philip dedicates his part to Uncle Wilfred Lawson (RAF/WW2/MIA)

Mountain View, California   B. M. E. de Silva and
www.newphysics.us   P. J. Lawson

© 2005 Dr. Brian M.E. de Silva & Philip J. Lawson. All Rights Reserved.

No part of this book may be reproduced, stored in a retrieval system, or transmitted by any means without the written permission of the author.

First published by AuthorHouse 12/19/05

ISBN: 1-4208-8975-3 (e)
ISBN: 1-4208-8769-6 (sc)

Printed in the United States of America
Bloomington, Indiana

This book is printed on acid-free paper.

Preface

This series of lectures on physical phenomenon is a brainchild of one of the authors (PJL) and is a culmination of several decades of nontraditional insights, into the study of post relativity physics.. After schooling in wartime London at the height of the blitz he went to Croydon Technical College (London) to study power and telecommunications engineering on a student/work program at the Mullard Research Laboratories in Salfords, Essex. While at the linear accelerator laboratory, he was assigned to work on a beam deflection assembly that scanned a 5 MeV beam 8 1/2 inches across a conveyor belt. The object was to irradiate sealed meat products inside a plastic wrap, for purposes of storage at room temperature without the need for cooking. It was also used to cross-link polymers for durable insulation at higher temperatures required for soldering insulated wires onto critical electronic assemblies.

The power signal generator and scanner that existed at the time was only able to achieve a ¾ inch deflection of the 5 MeV beam. As the resident engineering intern, he was asked to try his hand at making it achieve the required 8.5 inch deflection. He used 6 equations of the type used in electric motor design and added one more to allow for Einstein's relativistic mass increase, (which translated into an effective mass of approximately 8 times the rest mass) of the accelerated electrons. A large gap for the beam was machined from the laminations of a large 50 Hz transformer (required by his equation set ; to handle the magnetization level. This method was based on the assumption that the ferrite core used in the initial design had saturated at this level of magnetization and was the reason for the failure of the first scanner. The magnetic field used initially, was only able to deflect a lower voltage beam but not particles traveling at speeds close to the speed of light. The new deflection assembly gave the desired 8 1/2 inch deflection, and was exhibited at the 1957 National Physical Society Exhibition as an add on to the clinical accelerator normally used for cancer treatment thus increasing its market potential.

Philip then immigrated to Canada in 1957, where he worked for a number of years on classified military projects for the USAF After working at General Dynamics in Rochester NY, he became the Senior Electrical Engineer in the Metrology department at the Kennedy Space Center in Florida. The last 18 plus years of his professional career was spent at the Lockheed-Martin Missiles and Space Company in Sunnyvale, California contributing to several defense projects for the US Govt. This was followed by completion of several courses in modern physics at Stanford University.(Palo Alto) He brings a unique perspective into the world of space-time visualization, and his curiosity and spirit animates much of this monograph. His hands on engineering background has led him on a lifelong quest for practical models for visualizing the physical universe we inhabit and in the process has been forced to "walk the road less traveled".

The first author (BMEdeS) with his exposure to the rigors of the Cambridge University Mathematical Tripos has been able to embed these visual and empirical ideas into the structured language of higher mathematics. That is the only reason why he is the first author and not the other way around. He was reminded as he embarked on this collaborative effort , of a physics question his father had set for the London University Inter Science Examination in the late 1930s → "Discuss the statement that it was Faraday that made the name of Maxwell famous ".

Both authors intend to follow up this work, with a series of significantly less mathematically challenging books aimed at high school and junior college level readers. The present book is primarily aimed at final year mathematics and physics honors students and at mature researchers and practicing engineers with interests in theoretical physics and astronomy. It is also intended as a reference source for the later books.

The major topics covered in this presentation include the following:

- Relativistic effects of space-time and mass-velocity. We believe the primary difference between the absoluteness of Newtonian space and time, with their attendant independence, and the relativistic notions of space and time, is the mediating nature of the electromagnetic field described by Maxwell's equations.

- This intermediary nature of electromagnetism. is best illustrated by the light cone in 4 dimensions, or its 3 dimensional analogue, the light sphere. The physics is formulated within the framework of a flat 4 dimensional space-time in the absence of matter. The wave particle duality necessitated by the dictums of quantum mechnics, require the photons be smeared on the surface of the spherical wave front. This facilitates the transition from the geometry of Einstein to the 2 dimensional world of particles, whose trajectories reside on the surface of an ever expanding sphere. When matter resides in the interior, there would be a corresponding deflationary effect, slowing down the expansion.

- For an observer on the surface of the light sphere, the PRESENT corresponds to time = 0, while the interior PAST corresponds to time > 0, with the FUTURE represented by time < 0. The light sphere expands outwards at the speed of light.

- These philospohical ruminations have direct applicability to some of the fundamental questions of the early universe. In particular, the light sphere provides a bridge between the Cosmological Principle and a more anthromorphic view of the universe

- The mechanism for the conversion of gravitational energy into electromagnetic energy , may be described "jump conditions" , correponding to distortions to the geometry of space-time, consistent with the conservation of charge.

- The application of infinitie power series such as a Taylor series expansion to space-time derivatives for modeling atomic and cosmological events.

CONTENTS

Introduction

Chapter
/ Page

1 / 1   Wave - Particle Duality in Classical Gravitation Theory

2 / 11   Space-Time in Modern Physics (Light Sphere)

3 / 45   Particle Dynamics in Six Dimensional Space-Time

4 / 59   Ordered Space-Time Derivatives in Cosmology and Particle Physics

5 / 95   Ultra Relativistic Motion of Charged Particles

Questions and Answers (for above chapters)

Ch 1 / 111,

Ch 2 / 143,

Ch 3 / 153,

Ch 4 / 193

Introduction

" Verily, it is easier for a camel to pass through the eye of a needle than for a scientific man to pass through a door. And whether the door be a barn door or a church door it might be wiser that he should consent to be an ordinary man and walk in rather than wait till all the difficulties involved in a really scientific ingress are resolved."

Sir Arthur Eddington

Since the revolution in physics in the first half of the 20th century culminating in the birth of the theory of relativity and quantum mechanics, the concepts of ***past, present & future*** within the context of particle physics and cosmology have intrigued generations of theoretical physicists. As this series of essays show, we can re-interpret these notions through novel applications to demonstrate their centrality to modern physics. For example, in classical dynamics the notions of past and future are interchangeable, since Newton's Laws of motion are invariant under time reversals. Another example, would be Dirac's hole theory where a positron can be modeled in certain applications as an electron moving backward in time. In electrodynamics, the solutions to the inhomogeneous partial differential equations for the vector and scalar potentials in the presence of a charge and current distributions, involve both retarded and advanced potentials. In elementary applications, part of the complete solution is generally blocked out on the grounds of either physical inadmissibility or on causality considerations.

Mathematically, past and future events constitute 2 disjoint sets or closed systems in 4 dimensional space-time, with the present forming an impenetrable boundary. These domains are populated by particles and their antiparticles or in some instances by virtual particles. The light cone is the primary vehicle for our study. In particular, we focus on its projection in 3 dimensions → the light sphere which is more amenable for visualization of the physical processes. For example, we approach in a more heuristic manner the unification of gravitation and quantum mechanics in Chapter 1 through an investigation of the deformation of the light sphere in a strong gravitational field described by Newton's inverse square law of gravitation. The physics of the light sphere is described by Maxwell's electrodynamics in conjunction with quantum mechanics. This analysis results in particles and virtual or "ghost" particles.

The virtual particles experience a force of repulsion, with corresponding hyperbolic orbits, instead of Keplerian orbits for the particles. In Chapter 2, we explore further the mass~radiation imbalance in the early universe, within the context of a probabilistic light sphere consistent with Heisenberg's Uncertainty Principle. In contrast to Chapter 1, we consider a gravitational field described by the Schwarzschild metric of general relativity and its implications for Hubble's Law. In addition, we investigate the "lensing" effect of many particles consistent with the dictates of special relativity. In Chapter 3, we symmetrize both space and time, resulting in 3 temporal coordinates, 2 of which are pure imaginary. The motion of particles in this 6 dimensional space-time are formulated in some detail including elastic collisions of particles. Some of the inhabitants of this world include particles traveling with speeds greater than the speed of light. Again, we have the dichotomy of particles and virtual particles. Chapter 4 is our signature contribution which is essentially a theory of everything that has eluded physicists for decades. We approach the problem in a more intuitive manner, using the space-time derivates to assemble a representative sample of some of the landmark formulae of theoretical physics into a linear combination resembling a Taylor series approximation. Examples are drawn from celestial mechanics, atomic physics and relativity. The model has the flexibility to incrementally incorporate complexities through higher order terms. We believe this generalizes Dirac's Large Number Hypothesis, he proposed in the 1930s. Included is a prescription for generating new particles that have a specified mass-velocity relationship. Chapter 5 concludes with a study of charged particles moving at velocities approaching the speed of light. We give a detailed derivation of the damping associated with the radiation field.

Chapter 1   Wave - Particle Duality in Classical Gravitation Theory

> " The era of atomic energy can be an era of admirable progress, an era of a better and easier life. But it can also be an era of inexpiable strife, surpassing in extent and all the horrors of the past where, with the aid of terrifying means of destruction, humanity runs the risk of completely destroying itself."
>
> Louis de Broglie, Nobel Laureate, discoverer of the wave nature of Electrons

Abstract

This paper studies the distortions to a spherical light wave as it interacts with a gravitational field. The analysis is based on the assumption of a flat 4 dimensional space-time. The wave theory of Huygens is superimposed on the Newtonian corpuscular model of light as a stream of particles whose masses are subject to the gravitational laws of motion. The photons arise from the annihilation of electron-positron pairs and are described by Bose-Einstein statistics. The calculations are based on quantum statistical mechanics with a quantum mechanical expectation value for the Newtonian potential energy. Our approach uses a grand partition function to describe the physical characteristics of the resulting solutions. Two such solutions are generated, both of which are required for a complete description of the physics: (1) closed form wave fronts analogous to the Kepler orbits of Newtonian gravitational theory (2) open form "virtual" wave fronts analogous to Coulomb repulsion in electromagnetic theory, as for example in the Rutherford scattering of atomic nuclei. These solutions correspond to the ***retarded*** and ***advanced*** potentials of electromagnetic theory.

Introduction

A major focus of theoretical physics in the inter war years 1918 - 1939 was in establishing a bridge to unify the electromagnetic theories of Maxwell and Faraday with Einstein's theory of gravitation. An early attempt at this synthesis was the 5 dimensional geometry of Kaluza-Klein. Since then many theories of increasing complexity have been proposed, without much observational or experimental backing.

We adopt a more empirical approach in which a class of physical problems is considered, that are at the intersection of electromagnetism and gravitation. We believe the wave function of quantum mechanics is the natural vehicle for bridging these 2 diverse phenomena. It effectively addresses the particle aspects using the gravitational laws of Newton and the wave aspects using classical electrodynamics Specifically, we consider the destruction of an electron-positron pair. The resulting electromagnetic radiation in the form of gamma rays can be modeled either as a wave or a particle depending on the circumstances. Each photon corresponds to a particle of mass $2m_0$ for every electron-positron destruction, where m_0 is the rest mass of an electron (or positron). The particle impact is assumed to be at low velocities. The wave nature of the radiation is described by a wave function Ψ, which for a single electron-positron pair satisfies the relativistic Klein-Gordon equation,

$$\hbar^2 \lozenge^2 \psi = 4 m_0^2 c^2 \psi; \qquad \lozenge^2 = \nabla^2 - \frac{1}{c^2}\frac{\partial^2}{\partial t^2}$$

(1)

where c is the speed of light and $\hbar$ is Planck's constant divided by 2π, and $\lozenge^2$ is the 4 dimensional D'Alembertian operator.

This partial differential equation has a pair of solutions which are functions of both space (**r**) and time (t) and correspond to either outgoing or incoming waves that are amplitude modulated. In the particle picture, $\psi(\mathbf{r},t)$ corresponds to the probability of finding the photon at **r**,t in a 4 dimensional spacetime Our attempt at blending electromagnetism with gravitation is summarized in the flowchart below. In addition, to properly fold the study of electromagnetism into classical gravitational theory requires the following considerations.

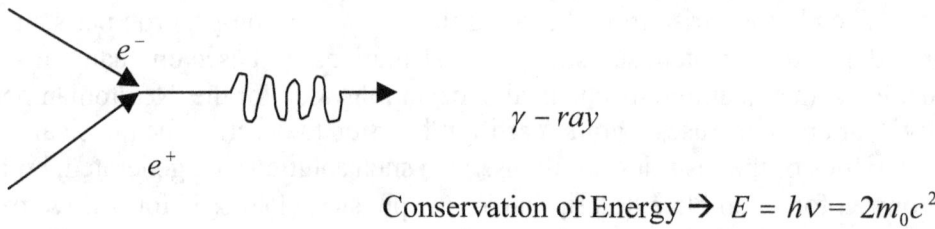

Conservation of Energy → $E = h\nu = 2 m_0 c^2$

- Quantum mechanics of many particle systems of electron-positron pairs
- Flat 4 dimensional space time - Newton's Gravitational theory is used instead of the Schwarzschild metric of general relativity
- Normalized wave function Ψ is constructed using the single particle wave functions $\{\psi\}$ consistent with Bose-Einstein Statistics for particles with zero spin. The $\{\psi\}$ satisfy the Klein-Gordon equation.
- Ψ is symmetric in the position coordinates $r_1, r_2, ..., r_n$ of the n particles
- Each photon is equivalent to a particle of mass $2 m_0$
- Newton's inverse square law of gravitation is used to calculate the mutual potential energy of the particles → may have possibilities for dark or "hidden" matter in some theories of cosmology.
- The concept of the "depletion" of gravitational mass is introduced in conjunction with the notion of present, past, and future events with respect to spherical waves. This has implications to stellar theory in which the inward pull of the gravitational forces is counteracted by the outward pressure caused by particles obeying Fermi-Dirac statistics. This usually manifests itself as a "slackening" in the gravitational inward pull on the star.
- Maxwell's equations of electrodynamics is the vehicle for studying the wave aspects of the theory of light

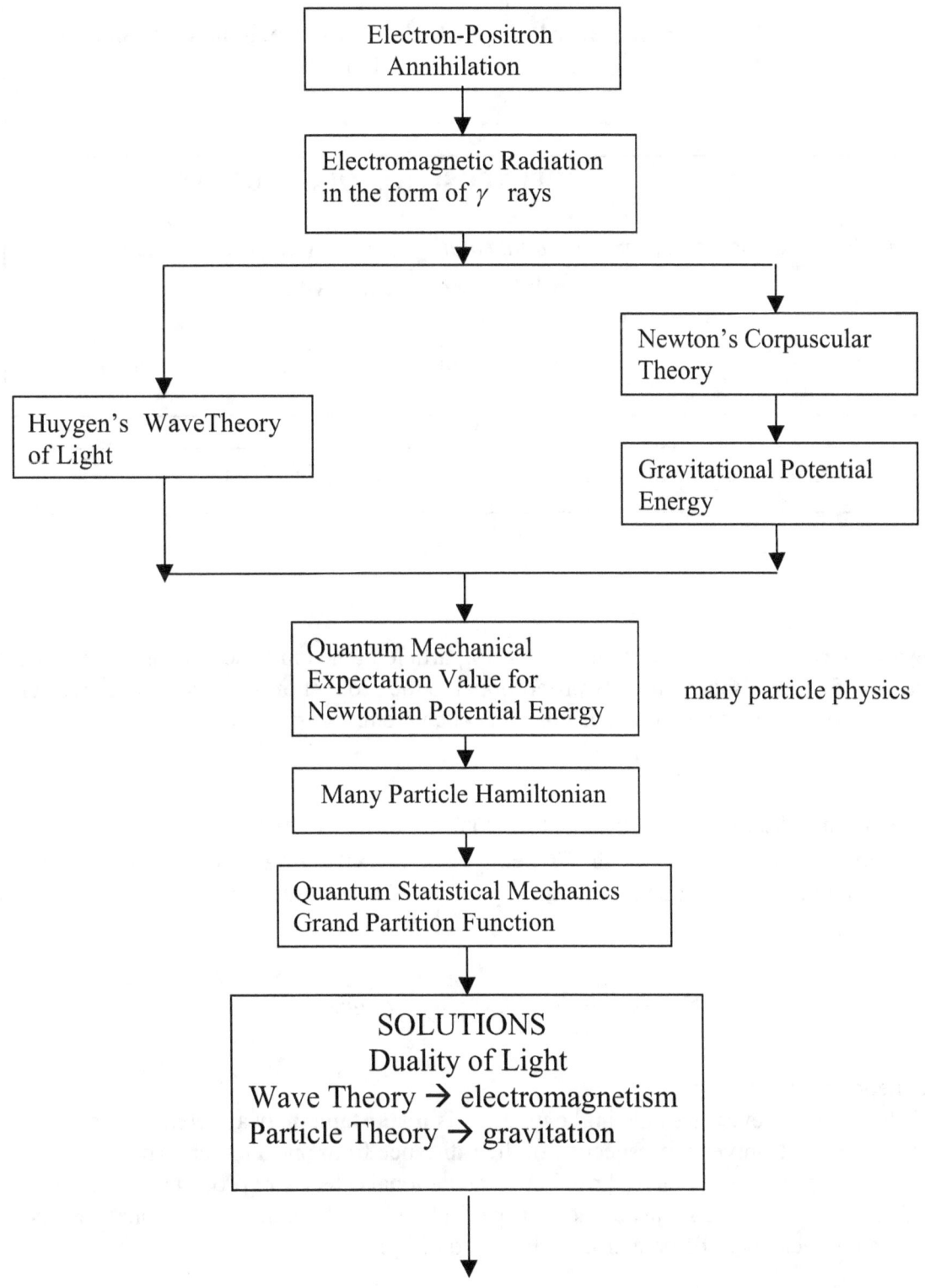

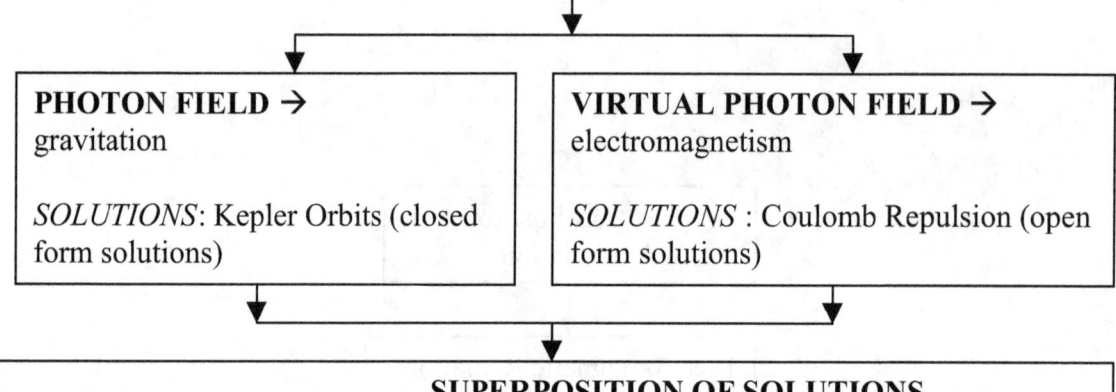

Gravitation

Newton's Law of Gravitation states that every particle in the universe attracts every other particle with a force that is directly proportional to the product of the masses and inversely proportional to the square of the distance between them.

$$F = G\frac{m_1 m_2}{r^2} \qquad (2)$$

where the particles of masses m_1, m_2 at a distance r apart will experience a mutual force of attraction of magnitude F, with G being the universal constant of Gravitation. This force is along the straight line joining the 2 particles. Their mutual potential energy is given by,

$$V = -G\frac{m_1 m_2}{r} \qquad ; \quad F = -gradV \qquad (3)$$

The theory implicitly assumes :
- The motion of every particle in the universe is **instantaneously** affected by every other particle in the universe irrespective of their distance from one another. This is sometimes known as Mach's Principle. Gravitational effects are propagated with an infinite velocity, in contrast to the theory of relativity, where the gravitational signals cannot travel at speeds greater than the speed of light.

- Particles interact in a 3 dimensional Euclidean or flat space. $ds^2 = dx^2 + dy^2 + dz^2$
- Space and Time are independent coordinates and have an absolute character

Newton's equations are a first order approximation to their relativistic counterparts when the velocity of motion is much smaller than the speed of light. The object of the research described in this paper is to examine the possibility of reconceptualizing Mach's Principle in relation to the 4 dimensional spacetime metric of special relativity.

$$ds^2 = c^2 dt^2 - dx^2 - dy^2 - dz^2 \qquad (4)$$

The light cone at any event $P_0 = (x,y,z,t)$ separates spacetime into a **timelike** ($ds^2 > 0$) and a **spacelike**
($ds^2 < 0$) domain. In a similar manner, the notion of **past** and **future** is introduced with respect to some defining event, which is taken to be the origin of 4-spacetime. For purposes of simplicity in this initial investigation, consider an electron and a positron colliding at $O = (0,0,0,0)$. The resulting gamma ray has an energy, $E = 2m_0 c^2$, where c = speed of light. In Huygen's theory of electromagnetic radiation, the resulting wavefronts correspond to concentric spheres centered at O with radius =ct. The photon resides on the surface of the sphere and corresponds to the **present**, while the **past** (spacelike) is the interior and the **future** the exterior (timelike) of the sphere. The position of the photon on the sphere is governed by the probability laws of quantum mechanics.
Consider distortions to the spherical wavefronts due to the destruction of a single pair of electron-positrons, each of rest mass m_0. More realistic simulations require a many particle approach that incorporates Newtonian gravitation within the framework of quantum statistical mechanics. This results in a reinterpretation of the action functional associated with Mach's Principle.

De Broglie Wavelength :

wave particle duality

$$\lambda = h/p \qquad (h = Planck's Constant) \qquad (5)$$

where λ is the wave length associated with a particle of rest mass $2m_0$ traveling with a velocity v and p is its linear momentum.

$$p = mass \times velocity \qquad (6)$$

In special relativity, *mass* is a function of *velocity* and is approximately equal to its rest mass at low velocities (in comparison to the velocity of light).

Wave Function

$\Psi = \Psi(\mathbf{r}, t)$; a complex function of position ($\mathbf{r}$) and time (t), corresponding to an event in 4 dimensional space-time. A simple qualitative definition is the following,
$|\Psi|^2 \, d\tau$ = Probability that the photon will be in a small element of volume $d\tau$ about the event ($\mathbf{r}$, t) in space-time.
Integration over the entire space-time must equal 1. For particles such as electrons or positrons, Ψ satisfies Dirac's equation, while for photons it must satisfy the Klein-Gordon equation. This is a consequence of a general theorem first derived by Pauli in 1940, concerning spin and statistics.

Electron-Positron Annihilation:

Consider an electron and a positron each of rest mass m_0 at the origin O of a coordinate system S at time $t = 0$. As a result of a slow collision a photon of energy E will be released in accordance with the law of conservation of energy m_0

$$E = h\nu = 2m_0 c^2 \qquad (7)$$

At a subsequent time t the photon (particle picture) will be on the surface a spherical wave front of radius = ct (wave picture). There is an equal probability that the photon will lie anywhere on the surface. The quantum mechanical calculations are therefore carried out with respect to surface integrals rather than volume integration. This is illustrated by the following simple example.

Bending of Huygen's Waves by a Massive Star

Consider a spherical wave front of radius ct centered on the origin O. The photon will be located somewhere on the surface of the sphere. The mass of the star is assumed to be M at a distance R from O. The mathematical steps are :

- The coordinates of the photon at any point P on the sphere. For simplicity we have used a spherical polar coordinate system at O. Calculate the distance of P from the star. Hence calculate the potential energy V of the photon in the gravitational field of the star.
- Calculate plane wave solutions of the photon wave equation. ψ will be a complex function of both position vector r and time t.
- Calculate the quantum mechanical expectation value of the potential energy $\langle V \rangle$ over the surface of the sphere. This is equivalent to "smearing" the photon over the surface, followed by an averaging process.
- Mathematical calculations show that

$$\langle V \rangle = +/- \, 2Gm_0 M / R \qquad t \leq R/c \qquad (8)$$

$$= +/- \, 2Gm_0 M / ct \qquad t \geq R/c \qquad (8a)$$

The expectation value $\langle V \rangle$ corresponds numerically to the classical gravitational potential energy of 2 particles of mass $2m_0$ and M a distance R apart. However, there are 2 possible representations because of the (+) and (-) signs. One photon trajectory corresponds to a "real" photon, while the other is a "ghost" or "virtual" trajectory. Both trajectories correspond to a particle of mass $2m_0$. The time R/c is the time taken for a light signal from the origin of the coordinate system to reach the star and is the critical time interface separating the two time zones. Because of the sign differences in $\langle V \rangle$ one corresponds to an attraction, and the other to a repulsion. The $1/r$ dependence of $\langle V \rangle$ implies the "real" trajectories are the Kepler orbits of Newtonian gravitational theory - circular orbits for $t < R/c$, and elliptical wave fronts for
$t > R/c$. While the "virtual" trajectories are analogous to the trajectories of two charged particles in a Coulomb repulsive field, such as in the case of hyperbolic paths for the Rutherford scattering of atomic nucleii for $t > R/c$.

Conclusions

For a single electron-positron pair, the time $t = R/c$ is the time taken for a light signal from the origin O, where the photon is created to reach the star. Before this time has elapsed, the star cannot distort the spherical wave fronts. This time has a direct analogy in electrodynamics, in which the "real" and "virtual or ghost" trajectories correspond to *retarded* and *advanced* potentials. In Part II, we consider the interaction of many electron-positron pairs. The trajectories generated in Part I, are subject to additional perturbations because of interaction effects.

Many body Solutions (Part II) = Unperturbed Solutions (Part I) + Perturbed Solutions due to Particle Interactions

The unperturbed solutions discussed in the present paper, would now be replaced by a particle of effective mass = $2m_0 N$, where N is the total number of electron-positron pairs. If N is sufficiently large, this could be a factor is some cosmological theories which postulate the presence of hitherto undetected forms of (dark) matter. If these particles satisfy Bose-Einstein statistics, they would have a repulsive or expansionary effect, manifesting itself in an apparent diminution of the gravitational tug.

The focus of this presentation is to demonstrate, the nexus between the phenomenon of electromagnetism embodied in the form of spherical waves and Newtonian gravitational theory as manifested through the wave particle duality of modern physics. For example, the source of the electromagnetic waves is the destruction of 2 particles : an electron and a positron. The physics is structured on a flat 4 dimensional space time In the particle picture, the photons are smeared on the surface of the waves, in accordance to the laws of quantum mechanics. This reveals an inner dichotomy between the 4 dimensional geometry of the wave picture and the 2 dimensional world of the particle model. The world lines of the photon reside on the surface of an ever expanding sphere. When matter resides in the interior of the sphere, there would be a corresponding deflationary effect, slowing down the expansion.

An alternate visualization, of the collapse of an electron and a positron into a point singularity, is the conversion of their mutual gravitational potential energy into electromagnetic energy. Geometrically, this process is accompanied by the 3 dimensional Euclidean space morphing into a 4 dimensional space-time. In these circumstances, one could speculate that conservation of energy demands a " jump condition" that addresses an energy deficiency in the conversion of gravitational energy into electromagnetic energy. Such a mechanism would be provided by distortions in the geometry of space time, consistent with charge conservation.

APPENDIX

This Section briefly reviews some of the basic mathematical equations introduced in the body of the text.

- Klein-Gordon Equation. In special relativity, the energy E of a particle of rest mass m_0 is related to its momentum p by the equation,

$$E = c\sqrt{p^2 + m_0^2 c^2} \tag{9}$$

$$p = \frac{m_0 v}{\sqrt{1 - v^2/c^2}} \quad ; \quad m = \frac{m_0}{\sqrt{1 - v^2/c^2}} \quad ; \quad p = mv \tag{9a}$$

where m is the mass of the particle and v is its velocity.

In quantum mechanics, E and p correspond to operators

$$E = i\hbar \partial/\partial t; \; p = -i\hbar \nabla \tag{9b}$$

operating on the wave function ψ, resulting in the Klein-Gordon partial differential equation. General
solutions are obtained using a separation of variables technique,

$$\psi(\mathbf{r}, t) = F(\mathbf{r})\, T(t); \qquad \mathbf{r},\, t = \text{independent variables} \tag{9c}$$

resulting in an ensemble of waves of the form,

$$\psi(\mathbf{r}, t) = \int \exp i[\mathbf{k}\cdot\mathbf{r} + c\sqrt{\frac{m_0^2 c^2}{\hbar^2} + k^2}\; t]\, d\mathbf{k} \tag{9d}$$

Simplified Solutions At time t, the photon will be on the surface of a sphere of radius $= ct$. Therefore, the photon wave function ψ must satisfy the normalization condition,

$$\int |\psi|^2 \, dS = 1 \tag{10}$$

the integration is over the surface of the sphere S. Since there is an equal probability of the photon lying anywhere on the surface, $|\psi|^2$ is a constant.

$$|\psi|^2 = 1/4\pi c^2 t^2 \; ; \; \psi = 1/(2ct\sqrt{\pi})\exp[i(vt - \mathbf{k}\cdot\mathbf{r})] \; ; \; \mathbf{k} = k\,\mathbf{n} \tag{10a}$$

where k is the constant wave number and $\mathbf{n}$ is a unit vector along the direction of propagation of the wave.

- The classical mean value of the potential energy V is given by,

$$\langle V \rangle = \frac{\int V\,dS}{\int dS} \quad \text{. the integration being over the surface of a sphere } S \text{ of radius} = ct. \tag{11}$$

Because of the mathematical form for ψ derived above for a single electron-positron pair, this mean value is equivalent to the quantum mechanical expectation value,

$$\langle V \rangle = \int \psi^* V \psi \, dS \; ; \; \psi\psi^* = |\psi|^2 = 1/4\pi c^2 t^2 = 1/\int dS \tag{11a}$$

For many particle systems, because of the interaction terms this equivalence is no longer valid.

- Calculation of V: using spherical polar coordinates centered at O, at any instant of time t, let the photon be at a point P, on the surface of a sphere of radius $= ct$.

Let,
$$P = (ct\sin\theta\cos\phi, ct\sin\theta\sin\phi, ct\cos\theta) \; ; \; A = (0,0,R) \tag{12}$$

where θ, ϕ are the polar angles, and A is the position of the star of mass M. Then the mutual gravitational potential energy is given by,

$$V = -G(2m_0)M/AP \qquad AP = \sqrt{c^2 t^2 - 2Rct\cos\theta + R^2} \tag{12a}$$

Substituting and simplifying gives,

$$\int V \, dS = -2Gm_0 M \iint \frac{c^2 t^2 \sin\theta}{\sqrt{c^2 t^2 - 2Rct\cos\theta + R^2}} \, d\theta d\phi \tag{12b}$$

The limits of integration are: $\theta = 0 \to \pi; \phi = 0 \to 2\pi$.

- The normalized many particle wave function is given by

$$\Psi(r_1, r_2, \ldots r_n, t) = \frac{1}{\sqrt{n!}} \sum_\rho \psi_{\rho(1)}(r_1, t) \psi_{\rho(2)}(r_2, t) \ldots \psi(r_n, t) \tag{13}$$

where $r_1, r_2, \ldots, r_n$ are the position vectors of the n particles, and the summation is over all permutations ρ of $\{1, 2, \ldots, n\}$. Ψ is symmetric in $r_1, r_2, \ldots, r_n$ because of the requirements of Bose-Einstein Statistics. The ψ's are the solutions of the Klein-Gordon Equation for the individual photons.

- The n-particle Hamiltonian

$$H = H(r_1, r_2, \ldots, r_n; p_1, p_2, \ldots, p_n) \tag{14}$$

where $p_1, p_2, \ldots, p_n$ are the respective particle momenta. In classical mechanics,

$$\frac{dr_i}{dt} = \frac{\partial H}{\partial p_i} \quad ; \quad \frac{dp_i}{dt} = -\frac{\partial H}{\partial r_i} \quad ; \quad i = 1, 2, \ldots, n \tag{14a}$$

In quantum mechanics,

$$[r_l p_m - p_m r_l] = i\hbar \delta_{lm} \tag{14b}$$

- Grand Partition Function,

$$\Xi = \exp(-H/kT) \tag{14c}$$

where k is Boltzmann's Constant and T is the absolute temperature.

Chapter 2 Space-Time in Modern Physics (Light Sphere)

> "Perhaps we all speak too soon and too glibly about space and time. They represent deep mysteries—metaphors for the very nature of being and existence. All the laws of science do not even begin to explain the mystery. They do, however, give us some insight and an important viewpoint, and they are certainly valuable."
>
> Roger Jones, Theoretical Physicist :

Abstract

This paper describes, within the framework of quasi-Newtonian physics, some features of space-time as it relates to the large scale properties of the universe. In particular, we focus on the transition from a radiation dominated universe immediately after the big bang to a matter filled "dark" universe. This process manifests itself in a red shifting of the electromagnetic spectrum as the universe ages. A modified Hubble's Law is used for the energy calculations, including an estimate for the total mass. The birth and death of the universe are singularities in the fabric of space-time and are examined using Heisenberg's uncertainty principle of quantum mechanics.

Introduction

There are three major cosmological theories: (1) big-bang → the universe and all the matter & radiation in it started from an event singularity and expanded outwards, forming in due course the observable universe (2) quasi-steady state theory of Hoyle-Narlikar → no beginning or end to the universe (3) Plasma cosmology → motion of charged particles, giving rise to galaxies. Computer simulations match closely the observed galactical shapes.
This paper is modeled on the big-bang theory of the universe It is thematically grouped into the following categories:

- Newtonian gravitation theory + special theory of relativity. The concept of a light cone is introduced to separate out PAST, PRESENT & FUTURE events in 4 dimensional space-time. For photons, the analog is the light sphere/probabilistic light sphere because of the wave-particle duality of light
- Conservation laws of physics are violated when there are singularities in the fabric of space-time → birth & death of the universe. Explanations based on quantum mechanics. Conservation laws apply to closed systems
- Hubble's Law applies to the motion of galaxies. A similar relationship is assumed at the early stages of the universe, prior to the formation of galaxies.
- Conservation of energy applied to the present epoch → radiant energy is converted into matter. Electron-Positron annihilation
- (probabilistic) light sphere → Doppler red shift (FUTURE) ; black body energy distribution(PRESENT); PAST inaccesible

1. Flat Space-time

For purposes of the present discussion, it is assumed the effects of gravity are negligible on the space-time structure of the universe. Therefore, space-time is conceptualized as flat with a metric,

$$ds^2 = c^2 dt^2 - dx^2 - dy^2 - dz^2 \tag{1}$$

in 4 dimensions. This is invariant under all Lorentz transformations of the coordinate system $Oxyzt$. Consider the motion of a point mass along a specified world line. A simple mathematical manipulation shows,

$$ds^2 = c^2 dt^2 \{1 - \frac{1}{c^2}[(\frac{dx}{dt})^2 + (\frac{dy}{dt})^2 + (\frac{dz}{dt})^2]\} = c^2 dt^2 (1 - v^2/c^2) \qquad (2)$$

where v is the velocity of the particle,

$$v^2 = (dx/dt)^2 + (dy/dt)^2 + (dz/dt)^2 \qquad (3)$$

2. Light cone

(i) if $ds^2 > 0$; $v < c$; → the separation of events is *time like*, because a simple time difference of events happen at the same place; $dx = dy = dz = 0$. Inside light cone → **future/past** events with respect to the origin, x=y=z=t=0

(ii) if $ds^2 < 0$; $v > c$; → the separation of events is *space like*, because of a spatial separation of simultaneous events $dt = 0$. Outside light cone. →the world of tachyons; **future /past** events have no meaning. This is an inaccessible zone.

(iii) if $ds^2 = 0$; $v = c$; → *null or light cone.* $c^2 dt^2 - dx^2 - dy^2 - dz^2 = 0$. (4)

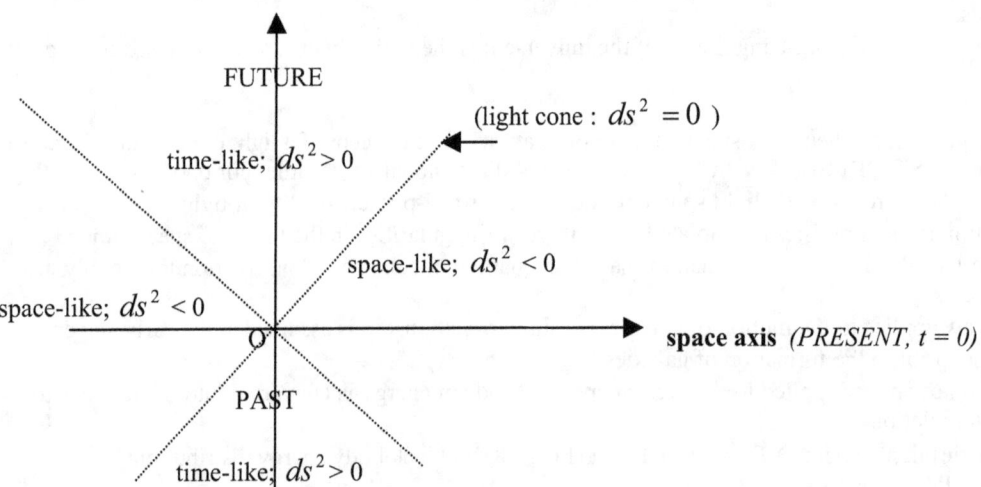

3. Light sphere

For particles travelling at the speed of light, $v = c$, $ds^2 = 0$. Such particles are photons. Therefore, from (1)

$$c^2 dt^2 - dx^2 - dy^2 - dz^2 = 0.$$

This implies (see Appendix),

$$c^2t^2 - x^2 - y^2 - z^2 = 0 . \quad \Rightarrow \quad x^2 + y^2 + z^2 = c^2t^2 \qquad (5)$$

The light cone in 4 dimensional space-time has collapsed into a sphere in 3 dimensions with center at the origin and radius = ct, the time $t.>0$ being a parameter labeling the concentric spheres. Because of the wave particle duality of quantum mechanics, this sphere must correspond to a light wave, described by the electromagnetism of Maxwell and Faraday. The photon will be smeared on the 2 dimensional surface of the 3 sphere. In this instance, it is possible to visualize a 4 dimensional object in the more intuitive 3 or 2 dimensions humans are accustomed to.

light cone → light sphere

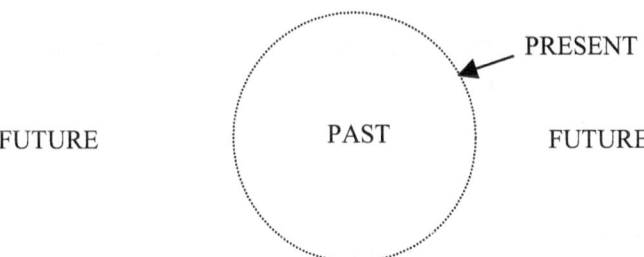

3. Closed Systems

Space-time with the above light-cone light-sphere dichotomy constitutes a closed system in which the intersection of the 2 sets,

$$\text{(time-like domain)} \cap \text{(space-like domain)} = \varnothing \qquad (6)$$

is the null set. The PAST and FUTURE domains in either 4 or 3 dimensions have no events in common. Because of the invariant character of ds^2, transitions between the 2 zones are forbidden, the set of events constituting the PRESENT forming an impenetrable barrier. The PAST is populated by Tachyons, "virtual" particles traveling at speeds greater than the speed of light.

4. Curved Space-time

The Schwarzschild metric associated with a mass M using spherical polar coordinates = (r, θ, ϕ) in 3 dimensional space is given by,

$$ds^2 = c^2 \gamma dt^2 - \frac{dr^2}{\gamma} - r^2 d\theta^2 - r^2 \sin^2 \theta d\phi^2 \qquad (7)$$

where,

$$\gamma = 1 - \frac{2MG}{rc^2} \; ; \; G = \text{universal constant of gravitation} \qquad (8)$$

The above metric has a singularity when $\gamma = 0$. This corresponds to,

$$1 - \frac{2MG}{rc^2} = 0 \quad \Rightarrow r = \frac{2MG}{c^2} \qquad (9)$$

This value of r, the radial distance from the origin is called the **Schwarzschild radius** for the mass M at O. When $\gamma = 1$, we have

$$ds^2 = c^2 dt^2 - dr^2 - r^2 d\theta^2 - r^2 \sin^2\theta d\phi^2 \tag{10}$$

But in 3 dimensions, the element of length in spherical polar coordinates (r, θ, ϕ) is,

$$ds^2_{(3)} = dr^2 + r^2 d\theta^2 + r^2 \sin^2\theta d\phi^2 = dx^2 + dy^2 + dz^2 \tag{11}$$

Therefore, the element of length in 4 dimensions is,

$$ds^2 = c^2 dt^2 - dx^2 - dy^2 - dz^2 \tag{12}$$

which corresponds to flat space-time

Flat space-time approximation: $\gamma \approx 1 \Rightarrow r >> 2MG/c^2$ or, in terms of the time taken to reach the Schwarzschild radius,

$$t >> 2MG/c^3 \tag{13}$$

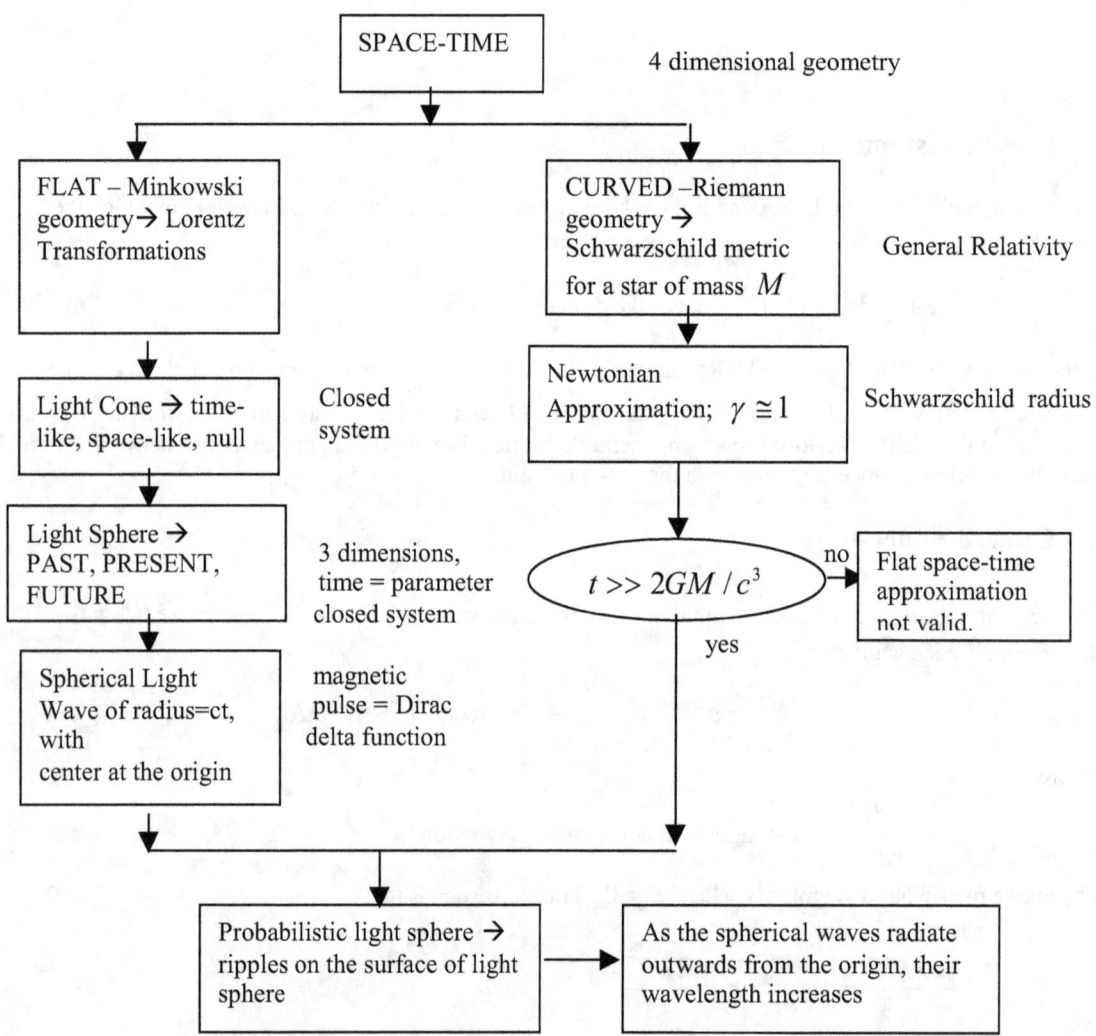

6. Probabilistic Light Sphere

Consider a massive particle of mass M at the center of a coordinate system $Oxyzt$. A single pulse of light originates from O at time $t = 0$. The light sphere at time $t \gg 2GM/c^3$ corresponding to the flat space-time approximation, is given by $x^2 + y^2 + z^2 = c^2 t^2$. By the wave-particle duality of light, the associated photon will travel on the surface of this sphere in closed Kepler orbits under the gravitational attraction of the mass M at the center. Because of the probabilistic nature of the orbits, the periphery of the sphere is serrated as shown below.

7. Doppler Effect

If ν is the photon frequency, then conservation of energy E gives

$$E = h\nu - \frac{GMh\nu}{c^2(ct)} = h\nu\left(1 - \frac{GM}{c^3 t}\right) \; ; \; h = \text{Planck's Constant}$$

where the photon is assumed to have an effective mass $h\nu/c^2$. The second term in the above expression corresponds to the gravitational potential energy of the photon at a distance ct from M.

As time t increases, since E is a constant, ν must decrease → i.e. the wave length of the radiation increases → spectrum is red shifted.

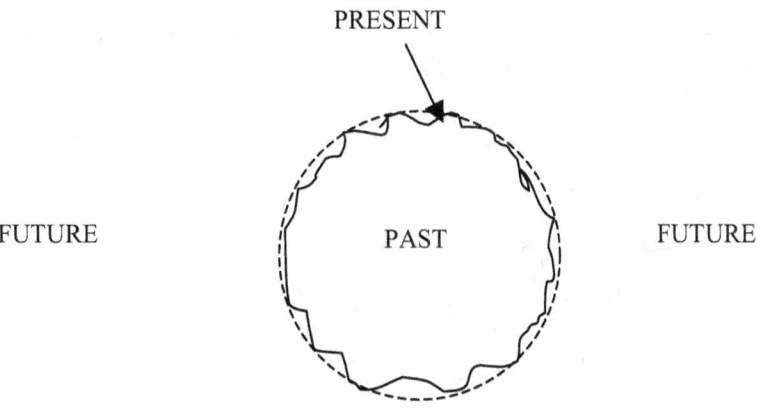

8. Gravitational Self Energy

Consider a uniform sphere of mass M and radius R, with its center at the origin O of the coordinate system. The gravitational field outside a spherical shell of radius r ($0 < r < R$) and thickness dr, will be equivalent to a point mass $\frac{4\pi r^3 \rho}{3}$, concentrated at its center O. Therefore the gravitational potential energy of the shell is given by,

$$-G\left(\frac{4\pi r^3 \rho}{3}\right)(4\pi r^2 dr\rho)/r$$

where $4\pi r^2 dr\rho$ is the mass of the shell, with the mass density ρ given by,

$$\rho = \frac{M}{4/3 \pi R^3}$$

Therefore, the gravitational potential energy is given by,

$$V = -\int G(4/3\pi r^3 \rho)(4\pi r^2 dr\rho)/r$$

with the limits of integration $r = 0 -> R$. On simplification this yields,

$$V = -\frac{16\pi^2 G\rho^2 R^5}{15} = -\frac{3GM^2}{5R} \qquad (14)$$

Therefore, the gravitational force is given by,

$$F = -\frac{dV}{dr}$$

$$= -\frac{3GM^2}{5R^2} \qquad (15)$$

in the **inward** radial direction. Therefore a uniform sphere will tend to **contract** under the influence of gravitation. The same tendency will prevail even if the mass density ρ is non uniform.

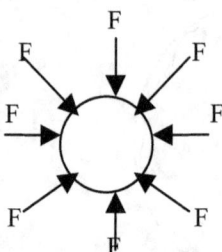

9. Expanding Universe

Suppose at time $t = 0$, all the matter comprising the universe was compacted into a point mass M. At this instant the big bang occurs, and the mass expands outwards as a sphere. Then the conservation of energy gives using (14),

$$E = \text{Kinetic Energy} + \text{Potential Energy} = \frac{1}{2}Mv^2 - \frac{3}{5R}GM^2 \qquad (16)$$

where v is the present rate of expansion of the universe, and R is its present radius. The expansion stops when $v = 0, R = \infty$. This gives $E = 0$.
Therefore,

$$v = \sqrt{\frac{6GM}{5R}} \qquad (17)$$

From modern astronomical observations and data incorporating modern technologies it is possible to estimate v and R. The above equation then enables us to estimate the total mass M in the universe.

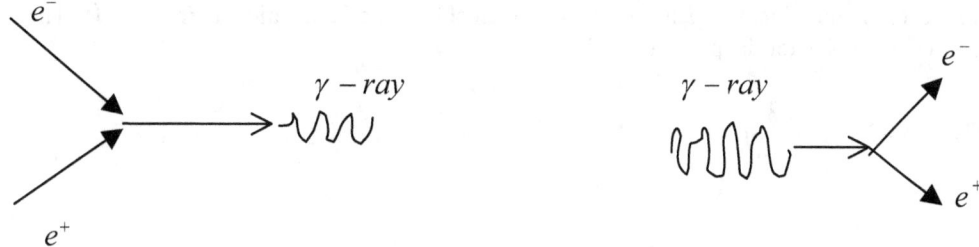

Matter into Radiation   Radiation into Matter

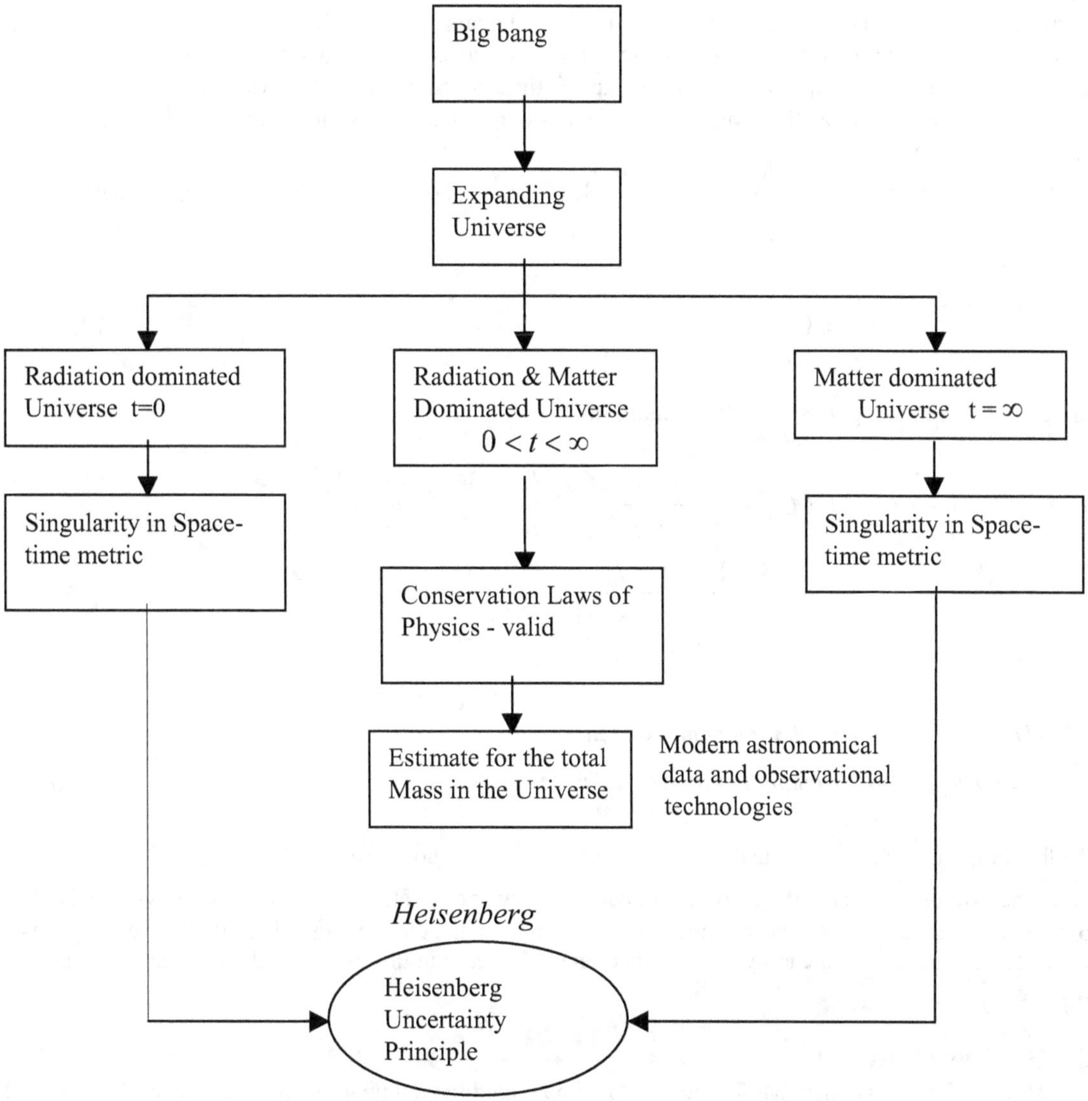

```
                    ┌───────────┐
                    │ Big bang  │
                    └─────┬─────┘
                          ↓
                    ┌───────────┐
                    │ Expanding │
                    │ Universe  │
                    └─────┬─────┘
        ┌─────────────────┼─────────────────┐
        ↓                 ↓                 ↓
┌───────────────┐ ┌───────────────┐ ┌───────────────┐
│ Radiation     │ │ Radiation &   │ │ Matter        │
│ dominated     │ │ Matter        │ │ dominated     │
│ Universe t=0  │ │ Dominated     │ │ Universe t=∞  │
│               │ │ Universe      │ │               │
│               │ │ 0 < t < ∞     │ │               │
└───────┬───────┘ └───────┬───────┘ └───────┬───────┘
        ↓                 ↓                 ↓
┌───────────────┐ ┌───────────────┐ ┌───────────────┐
│ Singularity   │ │ Conservation  │ │ Singularity   │
│ in Space-time │ │ Laws of       │ │ in Space-time │
│ metric        │ │ Physics-valid │ │ metric        │
└───────┬───────┘ └───────┬───────┘ └───────┬───────┘
        │                 ↓                 │
        │         ┌───────────────┐   Modern astronomical
        │         │ Estimate for  │   data and observational
        │         │ the total     │   technologies
        │         │ Mass in the   │
        │         │ Universe      │
        │         └───────────────┘
        │
                  Heisenberg
        │         ╭───────────╮         │
        └────────→│ Heisenberg │←────────┘
                  │ Uncertainty│
                  │ Principle  │
                  ╰───────────╯
```

17

10. Matter & Radiation

Suppose a fraction k of the initial mass M is converted into electromagnetic radiation at time, $t = 0$. Then the conservation of energy gives for the present epoch,

$$E = \frac{1}{2}(1-k)Mv^2 + kMc^2 - \frac{3}{5vt}G(1-k)^2 M^2 \; ; \qquad 0 \leq k \leq 1 \qquad (18)$$

where,

1 st term = kinetic energy of the mass $\rightarrow (1-k)M$
2 nd term = energy of the electromagnetic radiation
3 rd term = self energy of the gravitational field, as it expands outward at a uniform speed v.
It is assumed that the big bang occurs at $x = y = z = t = 0$. Immediately afterwards, matter expands radially outwards with a velocity $v(t)$, approaching the velocity of light c. It is further assumed that after the initial inflation of the universe measured in billions of years, the rate of expansion reaches a steady state constant value v, before it begins to decrease to a zero value as $t \rightarrow \infty$ and the size $\rightarrow \infty$, as sketched below. A more nuanced analysis, based on an explicit time dependence of v is briefly discussed below in the section on Hubble's Law. We believe, however the basic trends are still valid. For now, for simplicity it is assumed v is a constant.

As $t \rightarrow \infty, R = vt \rightarrow \infty, v \rightarrow 0 \Rightarrow E = kMc^2 > 0$. Therefore the total energy E in the universe is always positive. k is a maximum or minimum when

$$\frac{dk}{dt} = 0 \qquad (19)$$

Therefore differentiating (18) using the condition (19),

$$[-\frac{1}{2}Mv^2 + Mc^2 + \frac{6G(1-k)M^2}{5vt}]\frac{dk}{dt} + \frac{3G(1-k)^2 M^2}{5vt^2} = 0$$

$$\therefore \quad \frac{dk}{dt} = -\frac{3G(1-k)^2 M^2}{5vt^2[-\frac{1}{2}Mv^2 + Mc^2 + \frac{6G(1-k)M^2}{5vt}]} \leq 0 \qquad (20)$$

$\therefore t = 0 \rightarrow k_{max} = 1 \Rightarrow$ *radiation dominated universe* (21a)
$\therefore t = \infty \rightarrow k_{min} = 0 \Rightarrow$ *matter dominated universe* (21b)

The function $k(t)$ is a monotonic decreasing function of t as shown below.

Under the extraordinary conditions of the big bang, energy conservation only occurs for $t > 0$, while $t = 0$ corresponds to a singularity in space-time. As time increases, the electromagnetic (radiation) energy gets converted into gravitational energy with a concomitant decrease in the electric and magnetic fields in the universe.

11. Hubble's Law

In Section 10 it was assumed that for most of the time including and up to the present epoch the expansion rate of the universe is constant. We relax this assumption, by an appeal to Hubble's Law which was originally formulated to describe the motion of galaxies.

Every galaxy moves away from an observer with a velocity directly proportional to the distance of the galaxy from the observer. The motion is along the line of sight. We assume that in the immediate aftermath of the big bang a similar law applies to the motion of matter, with an appropriate Hubble function.

Velocity of expansion of Universe

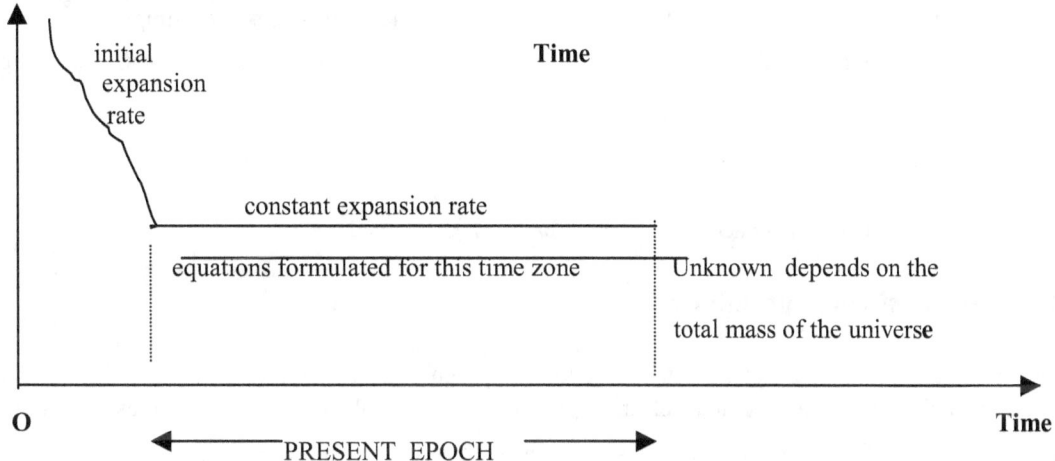

t = 0 instant of big bang

For non uniform expansion rates, the conservation of energy (18) gives (see Appendix)

$$E = \frac{3}{4}(1-k)Mv^2 + kMc^2 - \frac{3G(1-k)^2 M^2}{5r(t)} \qquad (21)$$

where we have used the approximation

$$R = r(t) = vt \qquad (22)$$

with v given by equation (17)

Fraction of Mass converted into radiation $k(t)$

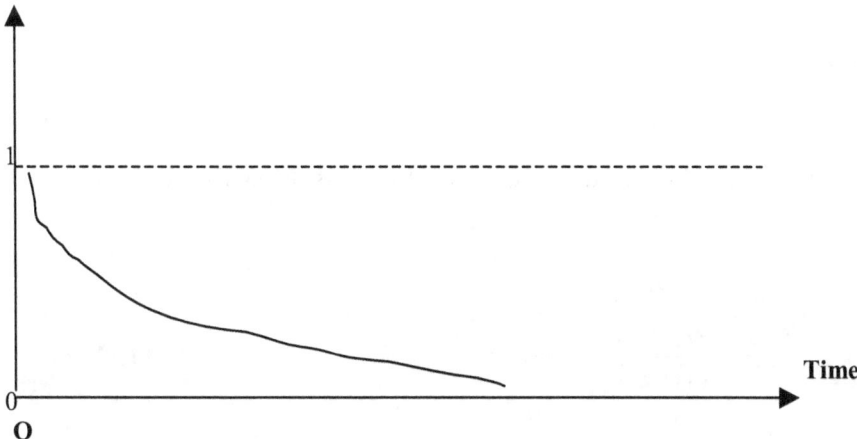

The velocity-distance relationship is now modeled on Hubble's law

$$v = v(t) = H(t)r(t) \; ; \; v = \frac{dr}{dt} \qquad (23)$$

where $H(t)$ is Hubble's function. Some of the early cosmological studies, were based on simple mathematical representations for $H(t)$ resulting in the Einstein, de Sitter, and Milne universes. Integrating,

$$r = r(t) = r_0 \, e^{\int_{t_0}^{t} H(t)dt} \; ; \; r_0 = r(t_0) \qquad (24)$$

Therefore for a given $H(t)$, a more accurate formula for $k(t)$ is obtained.

12. Electron-Positron Annihilation

We assume the process of converting mass into radiation is through the interaction of electron-positron pairs, and conversely, a photon is equivalent to an electron-positron pair. Each photon ($\gamma - rays$) thus created is equivalent to a particle of mass of mass $2m_0$, assuming the collisions occur at low velocities.

$$k(t)Mc^2 = 2m_0 c^2 n(t) \qquad (25)$$

where $n(t)$ is the number of photons. Since, $k(t)$ is a monotonic decreasing function, $n(t)$ decreases with time, as more radiant energy is converted into matter. The gravitational potential energy of each photon is

$$V = -\frac{6G(1-k)Mm_0}{5r} \qquad (26)$$

with a corresponding radial force of attraction given by,

$$F = -\frac{dV}{dr} = -\frac{6G(1-k)Mm_0}{5r^2} \qquad (27)$$

under this inverse square law, the photon will travel in closed Kepler orbits.

13. Heisenberg's Uncertainty Relations

$$\Delta r \Delta p \approx \hbar \qquad (28)$$

where $\Delta r, \Delta p$ are the uncertainties in position and momentum respectively. Assuming the momentum to be in the range ($p, p \pm \Delta p$),

$$\Delta p \approx hv(t)/c \qquad (29)$$

$$\therefore \quad \Delta r \approx \frac{c}{v(t)} = \lambda(t) \qquad (29a)$$

where $\lambda(t)$ is the wavelength of the radiation. As time increases, the frequency decreases and the wavelength increases, red shifting the electromagnetic spectrum As time increases, Δr increases and the photon trajectories become more diffuse or chaotic resulting in black body radiation.
The corresponding uncertainty relationship between Energy and time of occurrence is given by,

$$\Delta E \Delta t \approx \hbar \qquad (30)$$

This means that at the moment of the big bang $t = 0$, $\Delta t = 0 \Rightarrow \Delta E = \infty$. The discussion in Section 10 shows $E \to 0$, as $t \to \infty$. This implies if $\Delta E = 0 \Rightarrow \Delta t = \infty$ or if $\Delta t = 0 \Rightarrow \Delta E = \infty$. Therefore at the beginning of the universe, its total energy is indeterminate, while the end of the universe is characterized by either $E = 0$ in which case its end time is indeterminate or if the end time is known ($t = \infty$), then the energy is unknown. The beginning and end of the universe is characterized by singularities in the space-time structure and consequently the conservation laws would be inoperable.

Energy distribution E_λ

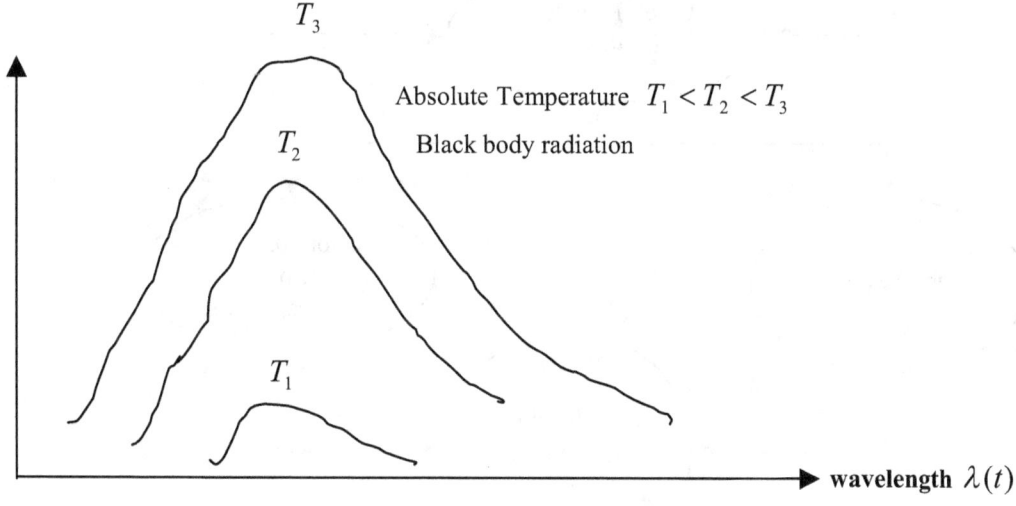

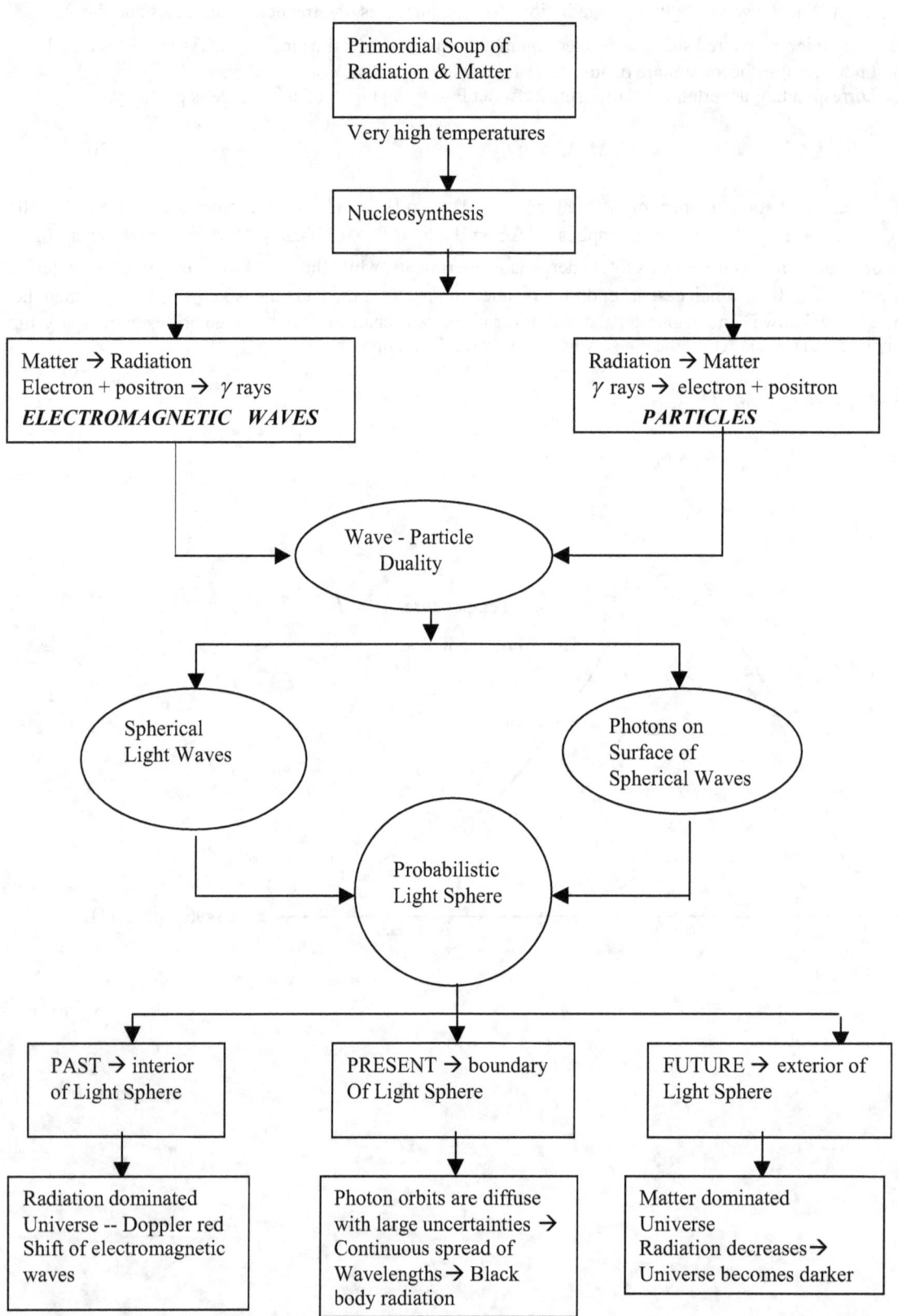

APPENDIX

- The kinetic energy of a sphere of mass M that is expanding radially outwards with a velocity analogous to Hubble's Law is given by,

$$\text{K.E.} = \int_0^r \frac{1}{2}(\rho dV)v^2 = \frac{1}{2}\int_0^r [\rho(r).r^2 \sin^2\theta dr d\theta d\phi] H^2 r^2$$

$$= \frac{1}{2} H^2(t) \int_0^r \rho(r).4\pi r^2 dr\, r^2 \quad (31)$$

Assuming a uniform mass density within a sphere of radius r $\quad \rho(r) = \dfrac{3M}{4\pi r^3}$, we have on simplification,

$$\text{K.E.} = \frac{3}{4} M H^2(t) r^2 = \frac{3}{4} M v^2 \quad (32)$$

- From Section 9 the maximum mass of the universe is given by (when $v = c$),

$$M_{max} = \frac{5c^2 R}{6G} \quad (33)$$

where R is the present value of the radius of the universe

- With the usual notation, the Lorentz transformation equations of special relativity are

$$x' = \frac{x - vt}{\sqrt{1 - \dfrac{v^2}{c^2}}} \quad (34a)$$

$$y' = y \quad (34b)$$

$$z' = z \quad (34c)$$

$$t' = \frac{t - \dfrac{vx}{c^2}}{\sqrt{1 - \dfrac{v^2}{c^2}}} \quad (34d)$$

where v is the velocity of the primed coordinate system relative to the unprimed system, along their common $Ox, O'x'$ axes. Initially, the 2 coordinate systems coincide, with $O \equiv O'$ and

$$x = y = z = t = 0 \iff x' = y' = z' = t' = 0 \tag{35}$$

Suppose a photon travels at the speed of light $v = c$, with the primed coordinate system moving with the photon. In other words, the origin O' is centered on the photon.

$$x' = 0 \implies x = vt \tag{36}$$

From (34d),

$$\therefore t' = \underset{v \to c}{Lt} \frac{t(1 - \frac{v^2}{c^2})}{\sqrt{1 - \frac{v^2}{c^2}}} = 0 \tag{37}$$

For example, within the interior of the light sphere,

$$x < ct$$

and from (34d),

$$t' = \frac{t - \frac{vx}{c^2}}{\sqrt{1 - \frac{v^2}{c^2}}} > \frac{t(1 - \frac{v}{c})}{\sqrt{1 - \frac{v^2}{c^2}}} > 0 \tag{37a}$$

while outside the light sphere,

$$x > ct$$

and from (34d),

$$t' = \frac{t - \frac{vx}{c^2}}{\sqrt{1 - \frac{v^2}{c^2}}} < \frac{t(1 - \frac{v}{c})}{\sqrt{1 - \frac{v^2}{c^2}}} < 0 \tag{37b}$$

Therefore, for an observer on the surface of the light sphere (PRESENT) $t' = 0$, while the interior (PAST) corresponds to $t' =$ positive, and the exterior (FUTURE) giving negative values for t'.

- Consider a point $P \equiv (x, y, z)$ on the surface of a light sphere of radius $= ct$ with center at the origin of coordinates in 3-space.

$$x^2 + y^2 + z^2 = c^2 t^2 \tag{38}$$

At time $= t + \delta t$, the point $P \to P' \equiv (x + \delta x, y + \delta y, z + \delta z)$ as the sphere expands radially outwards, with $PP' = c\delta t$

$$(PP')^2 = (c\delta t)^2 = (\delta x)^2 + (\delta y)^2 + (\delta z)^2$$

$$\therefore c^2 dt^2 - dx^2 - dy^2 - dz^2 = 0. \quad \Rightarrow ds^2 = 0 \tag{39}$$

Conversely, $ds^2 = 0 \Rightarrow c^2 t^2 - x^2 - y^2 - z^2 = 0$ \hfill (39a)

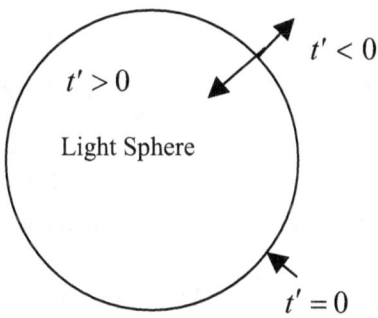

- At time t, the amount of matter in the universe is $(1-k)M$ and the Schwarzschild metric gives (Section 5),

$$\gamma(t) = 1 - \frac{2(1-k)MG}{rc^2} \tag{40}$$

at a distance r. As time increases, $k(t)$ decreases monotonically from 1 to 0 and $\gamma(t)$ decreases from 1 to $1 - \frac{2MG}{rc^2}$. Therefore, the space becomes more curved, due to the $1/\gamma$ term and the flow of time slows down.

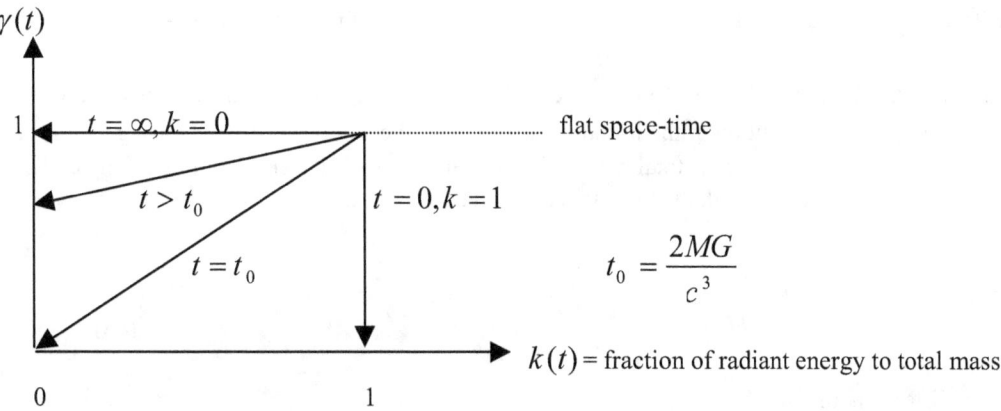

For $r = ct$,

$$\gamma(t) = \frac{2MG}{c^3 t} k(t) + 1 - \frac{2MG}{c^3 t} \qquad 0 \leq k \leq 1 \qquad (41)$$

This is a family of straight lines in $\gamma \approx k$, for values of t in $0 \leq t \leq \infty$ with slope and γ-intercept given by

$$\text{Slope (t)} = \frac{2MG}{c^3 t} = \frac{t_0}{t} \qquad (41a)$$

$$\gamma\_\text{intercept}(t) = 1 - \frac{2MG}{c^3 t} = 1 - \frac{t_0}{t} \qquad (41b)$$

All lines pass through the point (1,1). In Section 5 we saw that the flat space-time approximation is valid only if $\gamma(t) \approx 1$. From the above figure this requirement is satisfied if $t \gg t_0$

- strong gravitational fields: (a) $t \ll t_0$ or (b) $M \gg 1$ (c) both (a) & (b) true

Taylor series expansion of $k(t)$ about $t = 0$ gives,

$$\gamma(t) = \frac{2MG}{c^3 t} k(t) + 1 - \frac{2MG}{c^3 t} \qquad (42)$$

$$= \frac{2MG}{c^3 t} \left[k(0) + t k'(0) + \frac{t^2}{2!} k''(0) + \frac{t^3}{3!} k'''(0) + \ldots \right] + 1 - \frac{2MG}{c^3 t} \qquad (43)$$

$\because k(0) = 1$, $k'(0) = 0$ from equations (20, 21a), equation (43)

$$\gamma(t) = 1 + \frac{MGt k''(0)}{c^3} + O(t^2) \qquad (44)$$

where,

$k''(0) < 0$, since from equations (20, 21a), $t = 0$ is a point of maximum of $k(t)$.

A tentative sketch of the $\gamma(t)$ vs. *time* curve is shown below. A more accurate representation requires the numerical solution of the initial value problem of equations (20, 21a), using estimates for the present day values for the total mass M of the universe and its size R. For example, the intersection of the γ-curve with the time axis is given by equation (42),

$$\frac{t_0}{t} k(t) + 1 - \frac{t_0}{t} = 0$$

$$\therefore t = t_0 [1 - k(t)] < t_0 \qquad (45)$$

Again, differentiating (42)

$$\gamma'(t) = \frac{t_0}{t}[k'(t) + \frac{1-k(t)}{t}]$$

From L'Hopital's Rule,

$$\gamma'(0) = t_0[k''(0) - \frac{k''(0)}{2}] = -\frac{t_0 k''(0)}{2} < 0 \tag{46}$$

because from (20, 21a) $t = 0$ is a point of maximum for $k(t)$

The turning points of the curve are given by,

$$\gamma'(t) = 0 \tag{47}$$

There is at least one turning point, a point of minimum. The $\gamma(t)$ sketch gives a qualitative picture of the time intervals over which the flat space-time approximation $\gamma(t) \approx 1$ is valid.

- Perturbation expansions: zero order expansion → Consider the transformation,

$$\gamma(t) = 1 + t\gamma_1 + O(t^2) \; ; \; \gamma_1 = \frac{MGk''(0)}{c^3} \tag{48}$$

which in the immediate aftermath of the big-bang reduces the Schwarzschild metric of equation (7) to,

$$ds^2 = ds_{(0)}^2 + t\gamma_1(c^2 dt^2 + dr^2) + O(t^2) \tag{49}$$

where the first term on the right hand side corresponds to flat-space time, while the second term represents the first order perturbation to the Lorentzian space-time. In deriving the above result we have used the binomial expansion

$$(1 + t\gamma_1)^{-1} = 1 - t\gamma_1 + O(t^2) \quad \text{for } |t\gamma_1| \ll 1 \tag{50}$$

The second term in the metric represents the deviation from flat space-time, and the rate of this deviation depends on the initial rate of conversion of radiation into matter.

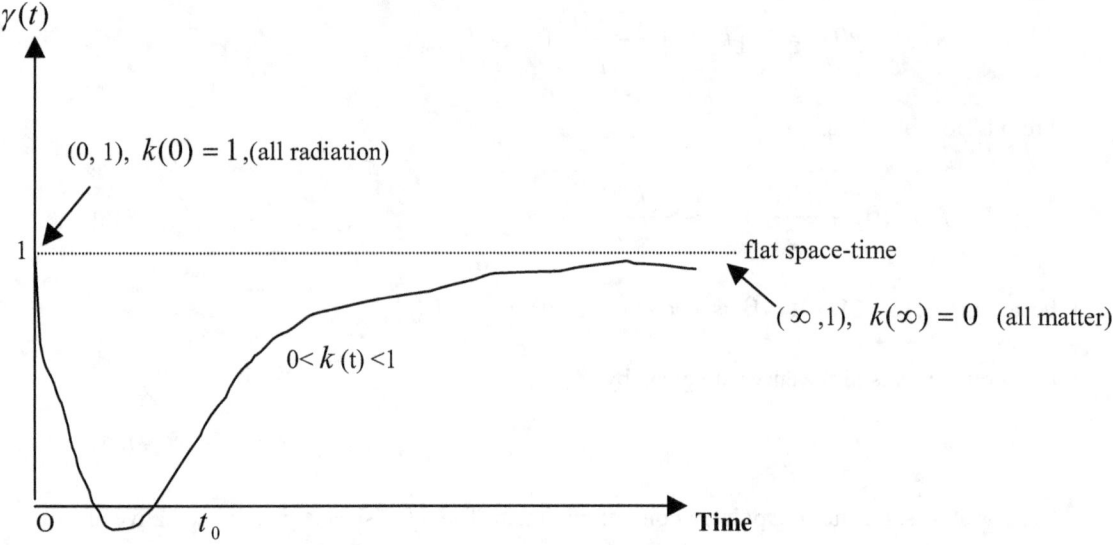

- A more accurate determination of the conversion factor $k(t)$ of Section 10 using Hubble's Law discussed in Section 11 is as follows:

$$E = \frac{(1-k)Mc^2}{\sqrt{1-\frac{v^2}{c^2}}} + kMc^2 - \frac{3}{5r}G(1-k)^2 M^2 \qquad (51)$$

with $k(0) = 1; k(\infty) = 0, r(\infty) = \infty, H(\infty) = 0 \Rightarrow v(\infty) = 0$ \qquad (52)

The first term in the energy expression is the relativistic energy of a sphere of mass $(1-k)M$ moving with a radial velocity v, the second term is the energy equivalent of a mass kM, while the third term is the gravitational self energy at a distance r from the site of the big-bang. Since the energy is a constant
The above boundary conditions ensure that the total energy $E = Mc^2$ as expected.
Re-arranging,

$$\frac{3GM^2}{5r}k^2 + \left(\frac{Mc^2}{\sqrt{1-\frac{v^2}{c^2}}} - Mc^2 - \frac{6GM^2}{5r}\right)k + \frac{3GM^2}{5r} + Mc^2 - \frac{Mc^2}{\sqrt{1-\frac{v^2}{c^2}}} = 0 \qquad (53)$$

where from equations (23, 24) on Hubble's Law

$$v = v(t) = H(t)r(t) \; ; \; v = \frac{dr}{dt}$$

$$r = r(t) = r_0 \, e^{\int_{t_0}^{t} H(t)dt} \; ; \; r_0 = r(t_0) \; ; \; t_0 = \text{present time}$$

The equation (53) in $k(t)$ is a quadratic, whose discriminant is given by,

$$\Delta = \left(\frac{Mc^2}{\sqrt{1-\frac{v^2}{c^2}}} - Mc^2 - \frac{6GM^2}{5r}\right)^2 - \frac{12GM^2}{5r}\left(\frac{3GM^2}{5r} + Mc^2 - \frac{Mc^2}{\sqrt{1-\frac{v^2}{c^2}}}\right)$$

$$= M^2 c^4 \left(\frac{1}{\sqrt{1-\frac{v^2}{c^2}}} - 1\right)^2 \quad > 0$$

$$\therefore \quad k(t) = \frac{-\left(\frac{Mc^2}{\sqrt{1-\frac{v^2}{c^2}}} - Mc^2 - \frac{6GM^2}{5r}\right) \pm \sqrt{\Delta}}{\frac{6GM^2}{5r}}$$

$$= 1 \quad \text{or} \quad 1 - \frac{5c^2 r(t)}{3GM}\left(\frac{1}{\sqrt{1-\frac{v^2}{c^2}}} - 1\right) \tag{54}$$

∴ The radiation to mass ratio is given by,

$$k(t) = 1 - \frac{5c^2 r(t)}{3GM}\left(\frac{1}{\sqrt{1-\frac{v^2}{c^2}}} - 1\right) = 1 - \frac{5c^2 r(t)}{3GM}\left(\frac{1}{\sqrt{1-\frac{H(t)^2 r^2(t)}{c^2}}} - 1\right) \tag{55}$$

- In Section 11, the graph of the velocity of expansion of the universe *vs* time, indicates that for time intervals significantly greater than the present era, velocity $\to 0$, as time $\to \infty$. However, some researches speculate that this velocity should in fact be increasing with time because of the presence of certain types of exotic matter/radiation. Our approach can easily handle this by a suitable choice of $H(t)$ *for sufficiently large values of time*.

  A simple example,

$$H(t) \approx t^{-\frac{2}{3}} \quad \text{for large } t \tag{56}$$

$$\therefore \quad r(t) \approx \exp(3t^{\frac{1}{3}})$$

$$\therefore \quad v(t) \approx t^{-\frac{2}{3}} \exp(3t^{\frac{1}{3}})$$

Velocity of expansion of universe

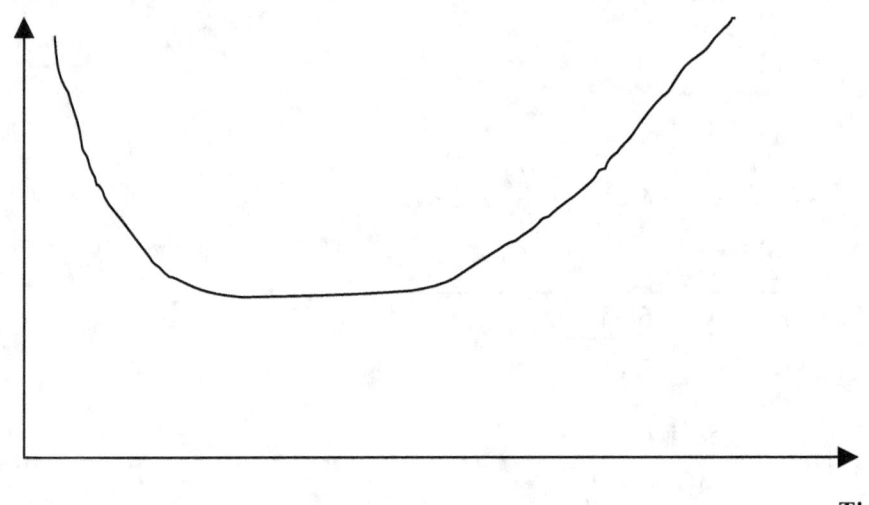

Time

The middle portion or "plateau" of this curve can be either stretched out or contracted by a suitable choice of a scaling factor, e.g. replacing $t \to t/t_*$. Alternatively, different Hubble functions $H(t)$ over different epochs can be constructed to model a required velocity distribution.

- A simple numerical technique for solving the initial value problem (IVP) of Section 10 is as follows:

  (i) divide the time interval of interest $[0,T]$ into n subintervals each length $\delta = T/n$ by points
  $: t_0, t_1, t_2, ..., t_{n-1}, t_n; \ 0 = t_0 < t_1 < t_2 < .... < t_{n-1} < t_n = T; \ t_r = t_0 + r\delta;$
  $r = 0,1,2,....,n;$

  (ii) $\dfrac{dk}{dt} = f(k,t) \ ; \ k(0) = 1$, from equation (20, 21a)

  $$f(k,t) = -\dfrac{3G(1-k)^2 M^2}{5vt^2[-\dfrac{1}{2}Mv^2 + Mc^2 + \dfrac{6G(1-k)M^2}{5vt}]}$$

  (iii) Euler integration using forward differencing yields

  $k_{r+1} = k_r + \delta f(k_r, t_r) \ ; \ r = 0,1,2,....,n-1$

  $k_r = k(t_r) \ ; r = 0,1,2,....,n-1, n$

(iv) generates $k_0, k_1, k_2, ..., k_{n-1}, k_n$ for values of time $= t_0, t_1, t_2, ..., t_{n-1}, t_n$

- Numerical Calculations : From equation(17) the present rate of expansion of the universe is given by

$$v = \sqrt{\frac{6GM}{5R}}$$

where, $M \approx 10^{56} \, gms = 10^{53} \, kg$
$R \approx 1.3*10^{10} \, lightyears = (3*10^{10}*3600*24*365)*1.3*10^{10} \, cm = 1.23*10^{28} \, cm$
$G \approx 6.67*10^{-11} \, Nm^2/kg^2$

Substituting,

$$v = 1.04*10^8 \, m/s \approx .3c$$

- Light cone in a gravitational field: From equation (7), $ds^2 = 0$

$$c^2 \gamma dt^2 - \frac{dr^2}{\gamma} - r^2 d\theta^2 - r^2 \sin^2 \theta d\phi^2 = 0$$

where from equation (8),

$$\gamma = 1 - \frac{2MG}{rc^2}$$

Consider a particular solution along a generator of the "curved" cone in 4 dimensions where,

$\theta = constant \; ; \; \phi = constant$

$$\therefore \frac{dr}{dt} = \pm c\gamma = \pm c(1 - \frac{2MG}{rc^2}) \tag{57}$$

Integrating,

$$(r - ct_0)^{ct_0} = \exp(\pm ct - r) \tag{58}$$

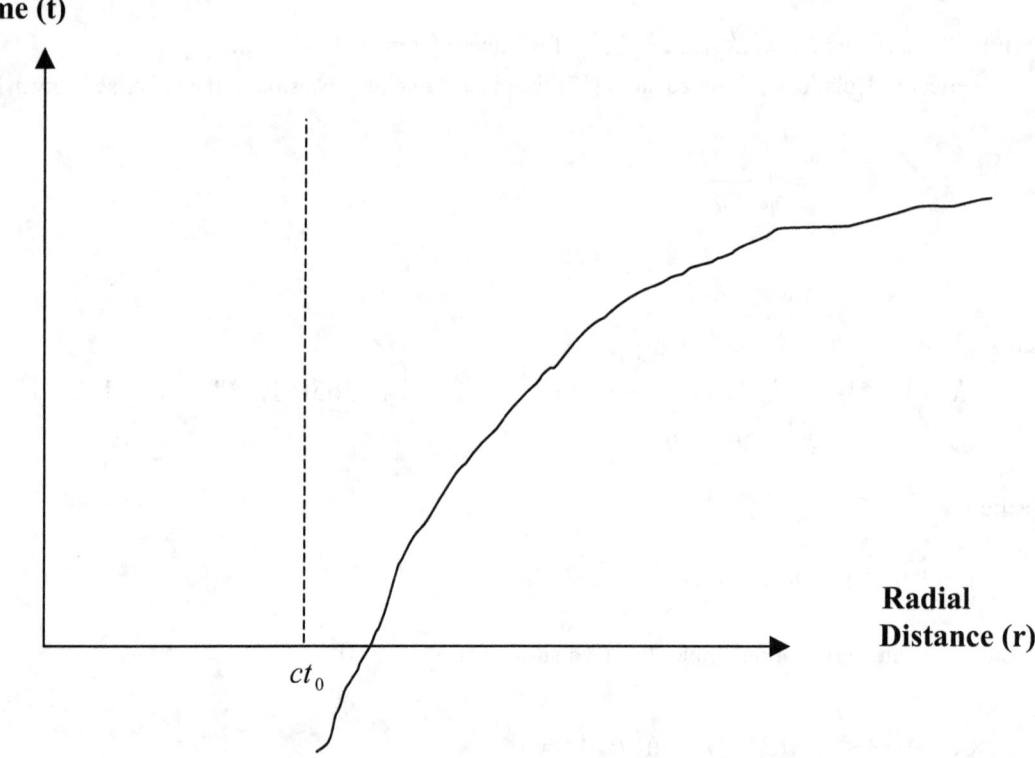

- A further example of an **closed system**

The force **F** acting on a particle of rest mass m_0 moving with a velocity **v** is given by,

$$\mathbf{F} = \frac{d}{dt}\mathbf{p} \tag{59}$$

Where **p** is the momentum,

$$\mathbf{p} = \frac{m_0}{\sqrt{1-\frac{v^2}{c^2}}}\mathbf{v} \tag{60}$$

Therefore the rate of work done by the applied force is,

$$\mathbf{F} \bullet \mathbf{v} = \mathbf{v} \bullet \frac{d}{dt}\left(\frac{m_0}{\sqrt{1-\frac{v^2}{c^2}}}\mathbf{v}\right) \tag{61}$$

But,

$$\frac{d}{dt}(\mathbf{v} \bullet \frac{m_0}{\sqrt{1-\frac{v^2}{c^2}}} \mathbf{v}) = \mathbf{v} \bullet \frac{d}{dt}(\frac{m_0}{\sqrt{1-\frac{v^2}{c^2}}} \mathbf{v}) + \frac{m_0}{\sqrt{1-\frac{v^2}{c^2}}} \mathbf{v} \bullet \frac{d}{dt} \mathbf{v} \qquad (61a)$$

and

$$\frac{d}{dt}(\mathbf{v} \bullet \frac{m_0}{\sqrt{1-\frac{v^2}{c^2}}} \mathbf{v}) = \frac{d}{dt}(\frac{m_0 v^2}{\sqrt{1-\frac{v^2}{c^2}}}) = \frac{m_0 v}{\left(1-\frac{v^2}{c^2}\right)^{\frac{3}{2}}}(2-\frac{v^2}{c^2})\frac{dv}{dt} \qquad (61b)$$

Therefore substituting (61a, 61b) in equation (61) gives,

$$\mathbf{F} \bullet \mathbf{v} = \frac{m_0 v}{\left(1-\frac{v^2}{c^2}\right)^{\frac{3}{2}}}(2-\frac{v^2}{c^2})\frac{dv}{dt} - \frac{m_0 v}{\sqrt{1-\frac{v^2}{c^2}}}\frac{dv}{dt} \qquad (62)$$

because,

$$\mathbf{v} \bullet \frac{d}{dt} \mathbf{v} = v \frac{dv}{dt}$$

$$\therefore \mathbf{F} \bullet \mathbf{v} = \frac{m_0 v}{\left(1-\frac{v^2}{c^2}\right)^{\frac{3}{2}}}\frac{dv}{dt} \qquad (63)$$

As $v \to c$, the work done in moving the point of application of the force a distance δr remains finite if

$$\frac{dv}{dt} \to 0$$

This condition is satisfied if $\delta v \to 0$, and $\delta t \to 0$. This latter requirement is satisfied due to the time dilation of the Lorentz transformation equations.

Therefore the total work done in reaching a velocity V is given by

$$\int_0^V \mathbf{F} \bullet \delta\mathbf{r} = \int_0^V \mathbf{F} \bullet \mathbf{v}\, dt = \int_0^V \frac{m_0 v}{\left(1-\frac{v^2}{c^2}\right)^{\frac{3}{2}}} \frac{dv}{dt}\, dt = -\frac{m_0 c^2}{\sqrt{1-\frac{V^2}{c^2}}} + m_0 c^2 \qquad (64)$$

If the force is conservative,

$$\mathbf{F} = -\nabla \Phi \qquad (65)$$

Then apart from an arbitrary additive constant,

$$\Phi = \frac{m_0 c^2}{\sqrt{1-\frac{V^2}{c^2}}} \qquad (66)$$

- From (59, 60),

$$\mathbf{F} = \frac{d}{dt}\left(\frac{m_0 \mathbf{v}}{\sqrt{1-\frac{v^2}{c^2}}}\right) = \frac{m_0}{\sqrt{1-\frac{v^2}{c^2}}} \frac{d}{dt}\mathbf{v} + \frac{m_0 v}{c^2\left(1-\frac{v^2}{c^2}\right)^{\frac{3}{2}}} \frac{dv}{dt}\mathbf{v} \qquad (67)$$

$$= \mathbf{F}_1 + \mathbf{F}_2$$

where,

$$\mathbf{F}_1 = \frac{m_0}{\sqrt{1-\frac{v^2}{c^2}}} \frac{d}{dt}\mathbf{v} \quad ; \mathbf{F}_2 = \frac{m_0 v}{c^2}\left(1-\frac{v^2}{c^2}\right)^{-\frac{3}{2}} \frac{dv}{dt}\mathbf{v} \qquad (68)$$

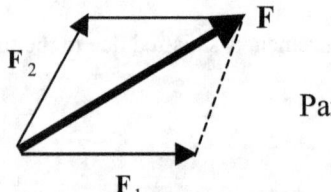

Parallelogram of forces

For particles traveling at speeds greater than the speed of light → tachyons in imaginary space-time,

$$v > c \qquad (69)$$

$$\mathbf{F} = = \mathbf{F}^*/i = (\mathbf{F}_1^* + \mathbf{F}_2^*)/i \tag{69a}$$

$$\mathbf{F}_1^* = \frac{m_0}{\sqrt{\frac{v^2}{c^2}-1}} \frac{d}{dt}\mathbf{v} \tag{69b}$$

$$\mathbf{F}_2^* = \frac{m_0 v}{c^2}\left(\frac{v^2}{c^2}-1\right)^{-\frac{3}{2}} \frac{dv}{dt}\mathbf{v} \tag{69c}$$

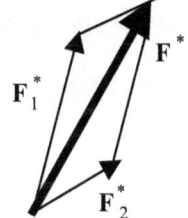   Parallelogram of forces for Tachyons

From the Lorentz transformation equations (34a-34d), we see,

$$dx' = \frac{dx - vdt}{\sqrt{1-\frac{v^2}{c^2}}} \tag{70a}$$

$$dt' = \frac{dt - \frac{vdx}{c^2}}{\sqrt{1-\frac{v^2}{c^2}}} \tag{70b}$$

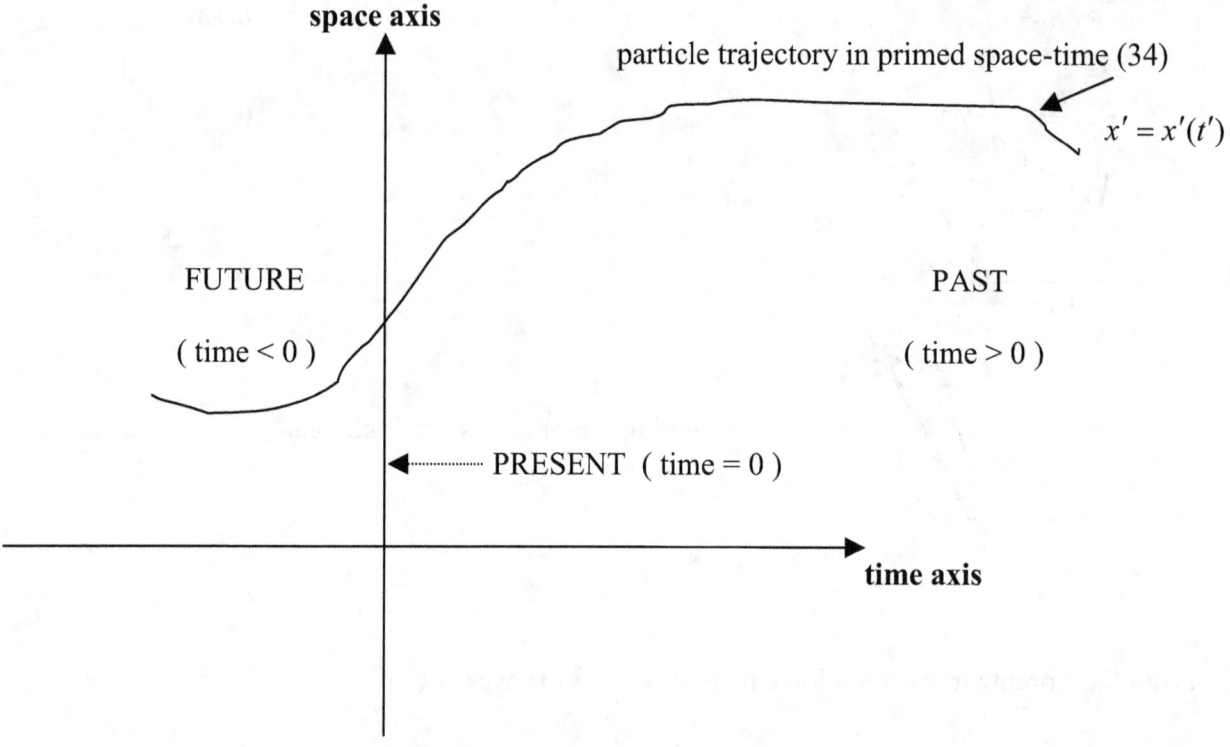

$$\therefore \quad \frac{dx'}{dt'} = \frac{\dfrac{dx}{dt} - v}{1 - \dfrac{v}{c^2}\dfrac{dx}{dt}} \tag{70b}$$

Therefore,

$$u' = \frac{u - v}{1 - \dfrac{v}{c^2}u} \tag{70c}$$

where,

$$u' = \frac{dx'}{dt'} \quad ; \quad u = \frac{dx}{dt} \tag{70d}$$

$u(t), u'(t')$ are the particle velocities in the primed and unprimed coordinate systems defined by the Lorentz transformation equations (34a-34d). As v varies, the curves (70c) all pass through the points (c, c), $(-c, -c)$. Therefore, these two points effectively act as the focii of a lensing effect similar to that encountered in geometrical optics.

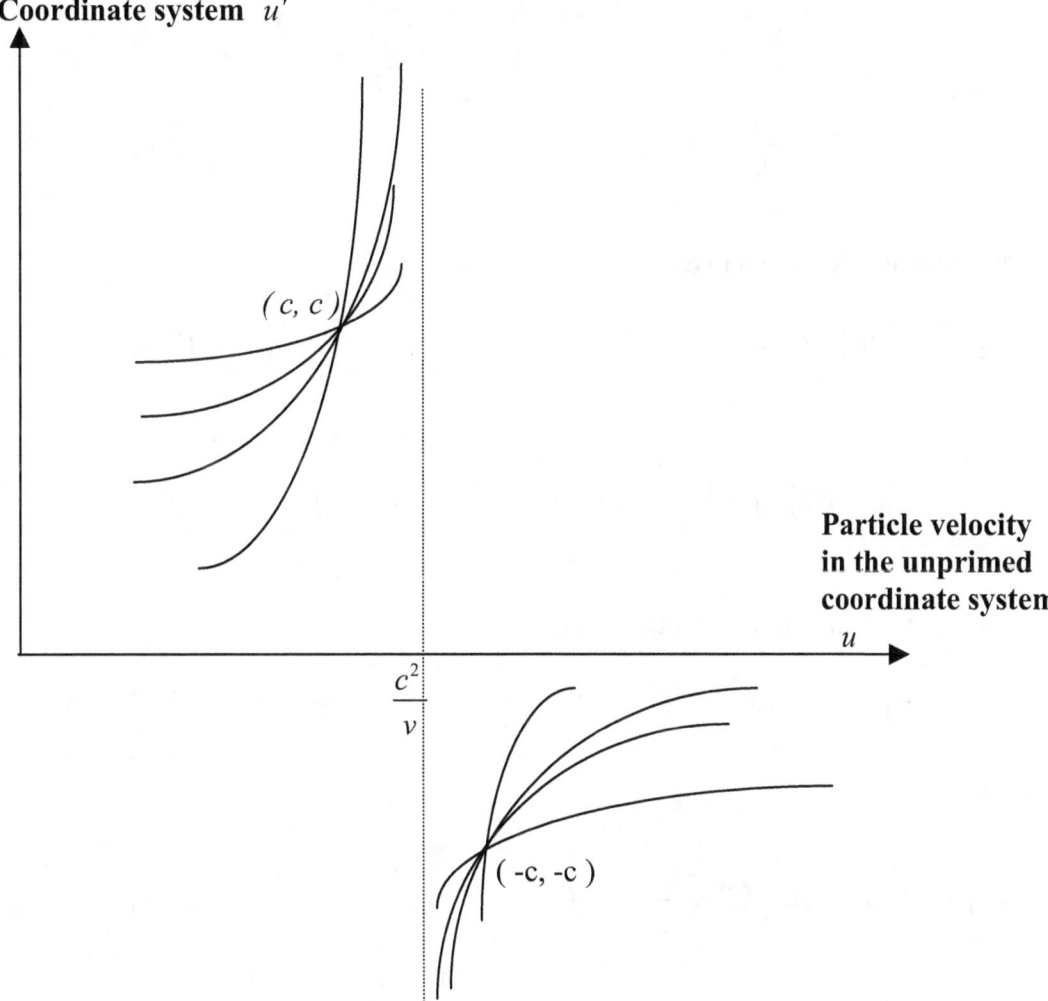

- For particles moving at speeds greater than the speed of light, $v > c$, Schrodinger's, equation becomes,

$$i\hbar \frac{\partial \Psi}{\partial t} = H\Psi \tag{71}$$

where the wave function Ψ is separated out into,

$$\Psi(\mathbf{r},t) = \psi(\mathbf{r})e^{-i\frac{Et}{\hbar}} \tag{71a}$$

with the energy E given by,

$$E = \frac{m_0 c^2}{\sqrt{1-\frac{v^2}{c^2}}} = \frac{m_0 c^2}{i\sqrt{\frac{v^2}{c^2}-1}} = \frac{m_0 c^2}{i} E^* \qquad (71b)$$

where, $\qquad E^* = \dfrac{m_0 c^2}{\sqrt{\dfrac{v^2}{c^2}-1}} \qquad (71c)$

The normalization condition gives,

$$\int |\Psi|^2 \, d\mathbf{r}\, dt = 1 \qquad (71d)$$

$$\therefore \quad \int |\psi(\mathbf{r})|^2 \, d\mathbf{r} \int e^{-\frac{2E^* t}{\hbar}} \, dt = 1$$

But the space normalization condition gives,

$$\int |\psi(\mathbf{r})|^2 \, d\mathbf{r} = 1$$

Therefore,

$$\int_0^\infty e^{-2\frac{E^* t}{\hbar}} dt = 1 \Rightarrow E^* = \frac{\hbar}{2} \qquad (71e)$$

$$\therefore \quad E^* \approx 5 \times 10^{-28} \quad \text{ergs} \qquad (72)$$

This is equivalent to a particle of mass,

$$m^* = \frac{E^*}{c^2} \approx 5.56 \times 10^{-49} \quad gm \qquad (72a)$$

This is about 10^{-22} times an electron mass, and qualifies possibly as a neutrino. Incidentally, the energy should be either ± because of the square root, corresponding to a neutrino or antineutrino. In the latter case the limits of integration should be from $[-\infty, 0]$ as an antineutrino is a neutrino going backward in time.
The lifetime of these particles can be estimated using the Heisenberg Uncertainty Principle.

$$\Delta E^* \Delta t \approx \hbar \qquad (73)$$

Therefore from (72),

$$\Delta t \approx 2 \text{ sec} \tag{73}$$

With such a short life span this particle probably cannot qualify as a neutrino.

- Rotation of coordinate axes → motion of charged particles. The Lagrangian for a particle of charge q in an electromagnetic field characterized by a scalar potential ϕ and a vector potential A is given by,

$$L = -m_0 c^2 \sqrt{1 - \frac{v^2}{c^2}} - q\phi + \frac{q}{c} A \bullet v \tag{74}$$

$$\approx \frac{1}{2} m_0 v^2 - q\phi + \frac{q}{c} A \bullet v \tag{74a}$$

Therefore the generalized momenta are given by,

$$p = \frac{\partial L}{\partial v} = \frac{m_0}{\sqrt{1 - \frac{v^2}{c^2}}} v + \frac{q}{c} A \tag{74b}$$

Therefore, the particle Hamiltonian is given by,

$$H = p \bullet v - L = \frac{m_0 c^2}{\sqrt{1 - \frac{v^2}{c^2}}} + q\phi = c\sqrt{m_0^2 c^2 + p^2} + q\phi \tag{74c}$$

Consider a coordinate system that is rotating with an angular velocity ω. If $\mathbf{v}_r$ is the velocity of the particle measured in the rotating system, the velocity in the fixed frame of reference is given by,

$$\mathbf{v} = \mathbf{v}_r + \omega \wedge \mathbf{r} \tag{74d}$$

Substituting (74d) in (74) gives,

$$L = -m_0 c^2 \sqrt{1 - \frac{(v_r + \omega \wedge r)^2}{c^2}} - q\phi + \frac{q}{c} A \bullet (\mathbf{v}_r + \omega \wedge \mathbf{r}) \tag{74e}$$

Using the approximation (74a), equation (74e) reduces to,

$$L = \frac{m_0}{2}(\mathbf{v}_r + \omega \wedge \mathbf{r})^2 - q\phi + \frac{q}{c}A \bullet (\mathbf{v}_r + \omega \wedge \mathbf{r})$$

$$= \frac{m_0}{2}\mathbf{v}_r^2 + m_0 \mathbf{v}_r \bullet (\omega \wedge \mathbf{r}) + \frac{m_0}{2}(\omega \wedge \mathbf{r})^2 - q\phi + \frac{q}{c}A \bullet \mathbf{v}_r + \frac{q}{c}A \bullet (\omega \wedge \mathbf{r})$$

Let us choose ω such that,

$$m_0(\omega \wedge \mathbf{r}) = -\frac{q}{c}A \qquad (74f)$$

$$L = \frac{m_0}{2}\mathbf{v}_r^2 - q\phi + \frac{m_0}{2}(\omega \wedge \mathbf{r})^2 - m_0(\omega \wedge \mathbf{r})^2$$

$$L = \frac{m_0}{2}\mathbf{v}_r^2 - q\phi - \frac{m_0}{2}(\omega \wedge \mathbf{r})^2 = \frac{m_0}{2}\mathbf{v}_r^2 - q\phi - \frac{q^2 A^2}{2m_0 c^2} \qquad (74g)$$

For weak magnetic fields A is assumed small and (74g) reduces to,

$$L \approx \frac{m_0}{2}\mathbf{v}_r^2 - q\phi \qquad (74h)$$

which is the classical Lagrangian for a particle in a purely electrostatic field. The effect of the magnetic field has been cancelled out. From Maxwell's equations,

$$\nabla \bullet \mathbf{B} = 0 \quad \Rightarrow \quad \mathbf{B} = \nabla \wedge A \qquad (74i)$$

Take,

$$A = -\frac{1}{2}(r \wedge \mathbf{B}) \qquad (74j)$$

with the magnetic field **B** assumed independent of position vector *r*. Therefore from elementary vector analysis we have the identity,

$$\mathbf{B} = -\frac{1}{2}\nabla \wedge (r \wedge \mathbf{B}) = -\frac{1}{2}[(\mathbf{B} \bullet \nabla)r - \mathbf{B}(\nabla \bullet r) - (r \bullet \nabla)\mathbf{B} + r(\nabla \bullet \mathbf{B})]$$

From (74f, 74j),

$$m_0(\omega \wedge \mathbf{r}) = -\frac{q}{c}A = \frac{q}{2c}(r \wedge \mathbf{B})$$

Therefore,

$$\omega = -\frac{q}{2m_0 c} \mathbf{B} \qquad (75)$$

This is the Larmor frequency. Without loss of generality let us assume the electric field **E** is along the *Ox* axis, while the magnetic field **B** is parallel to the *Oz* axis.

Let, $\mathbf{E} = E_0 \cos\Omega t \; \mathbf{i} \; ; \quad \mathbf{B} = B_0 \sin\Omega t \; \mathbf{k} \; ; \qquad (76)$

where (**i, j, k**) are unit vectors along the *Ox, Oy, Oz* axes respectively and E_0, B_0 are constants. Then in the rotating frame of reference,

$$m_o \frac{d}{dt} \mathbf{v}_r = q \mathbf{E} = q E_0 \cos\Omega t \; \mathbf{i}$$

$$\therefore \mathbf{v}_r = \frac{qE_0}{m_0 \Omega} \sin\Omega t \; \mathbf{i} \qquad (77)$$

where the constant of integration is selected so that $\mathbf{v}_r = 0$ at $t = 0$.

From (74d, 75),

$$\frac{d}{dt}\mathbf{r} = \frac{qE_0}{m_0\Omega}\sin\Omega t \; \mathbf{i} - \frac{qB_0}{2m_0 c}\sin\Omega t \; \mathbf{k} \wedge \mathbf{r} \qquad (78)$$

In component form,

$$\dot{x} = \frac{q}{m_0}(\frac{E_0}{\Omega} + \frac{B_0}{2c}y)\sin\Omega t \qquad (79a)$$

$$\dot{y} = -\frac{qB_0}{2m_0 c} x \sin\Omega t \qquad (79b)$$

$$\dot{z} = 0 \quad \Rightarrow z = cons\tan t = 0 \qquad (79c)$$

where without loss of generality we have taken the constant of integration to be zero. Write,

$$\xi = x + iy \qquad (80)$$

Then (79a, 79b) can be combined as,

$$\dot{\xi} = \frac{q}{m_0}(\frac{E_0}{\Omega} - \frac{iB_0}{2c}\xi)\sin\Omega t \qquad (81)$$

This reduces to,

$$\dot{\xi} + \frac{iqB_0\sin\Omega t}{2m_0 c}\xi = \frac{qE_0}{m_0\Omega}\sin\Omega t \qquad (81a)$$

Integrating factor $= \exp[\int \frac{iB_0 q}{2m_0 c}\sin(\Omega t)dt] = \exp(-\frac{iB_0 q}{2m_0 c\Omega}\cos\Omega t) \qquad (81b)$

$\therefore$ using (81b) in (81a), in accordance with the usual rules for handling this differential equation,

$$\frac{d}{dt}\left(\xi e^{-\frac{iB_0 q}{2m_0 c\Omega}\cos\Omega t}\right) = \frac{qE_0}{m_0\Omega}\sin\Omega t \ \exp(-\frac{iB_0 q}{2m_0 c\Omega}\cos\Omega t)$$

Integrating,

$$\xi = \frac{2E_0 c}{iB_0\Omega}(1 - e^{i\frac{B_0 q}{2m_0 c\Omega}\cos\Omega t}) \qquad (81c)$$

with the initial condition x=y=0 $\Rightarrow \xi = 0$ at t= 0. Equation (81c) simplifies to,

$$\xi = -\frac{2E_0 c}{B_0\Omega}\left\{-\sin\left(\frac{B_0 q}{2m_0 c\Omega}\cos\Omega t\right) + i\left[1 - \cos\left(\frac{B_0 q}{2m_0 c\Omega}\cos\Omega t\right)\right]\right\}$$

Equating real and imaginary parts yield,

$$x = \frac{2E_0 c}{B_0\Omega}\sin\left(\frac{B_0 q}{2m_0 c\Omega}\cos\Omega t\right) \qquad (82a)$$

$$y = -\frac{2E_0 c}{B_0 \Omega}\left[1 - \cos\left(\frac{B_0 q}{2m_0 c\Omega}\cos\Omega t\right)\right] \tag{82b}$$

Eliminating the time between equations (82a, 82b) we have,

$$x^2 + \left(y + \frac{2E_0 c}{B_0 \Omega}\right)^2 = \left(\frac{2E_0 c}{B_0 \Omega}\right)^2 \tag{82c}$$

This is the equation to a circles in the (x, y) plane with center $=(0, -\frac{2E_0 c}{B_0 \Omega})$ touching the Ox axis at the origin, for all values of $\frac{q}{m_0}$.

Let,

$$\vartheta = \frac{B_0 q}{2m_0 c\Omega}\cos\Omega t \quad ; \quad R = \frac{2E_0 c}{B_0 \Omega} \tag{83}$$

The radial and tangential equations of motion for the circular orbits of radius R given by (82a, 82b) are,

$$m_0 R \dot{\vartheta}^2 = -\frac{q}{c}(R\dot{\vartheta})B_0 \sin\Omega t + qE_0 \cos\Omega t \sin\vartheta \tag{83a}$$

$$m_0 R \ddot{\vartheta} = qE_0 \cos\Omega t \sin\vartheta \tag{83b}$$

In deriving the above equations we have used the Lorentz expression for the force,

$$\mathbf{F} = q(\mathbf{E} + \frac{1}{c}\mathbf{v} \wedge \mathbf{B}) \tag{83c}$$

For consistency, we have from (83) & (83b),

$$\ddot{\vartheta} = \frac{qE_0}{m_0 R}\cos\Omega t \sin\vartheta = -\frac{B_0 q\Omega}{2m_0 c}\cos\Omega t$$

$$\therefore \quad \sin\vartheta = -1 \tag{84}$$

$$\vartheta = -\frac{\pi}{2} \qquad (84a)$$

From (83a, 83b) using (84, 84a)

$$\cos \Omega t = 0 \qquad (85)$$

$$\therefore \quad t = \frac{(2n+1)\pi}{2\Omega} \qquad n = 0,1,2,3,..... \qquad (86)$$

This means the time is quantized. The particle lies on the circle (82c) at only discrete times given by (86). The trajectories between these time intervals are governed by the Heisenberg Uncertainty Principle. The uncertainty in time is given by,

$$\Delta t \approx \frac{\pi}{\Omega} \qquad (87a)$$

Therefore the uncertainty in energy is,

$$\Delta E \approx \hbar \Omega / \pi \qquad (87b)$$

This means that the energy spread increases with the frequency of the input electrical and magnetic fields.

Chapter 3   Particle Dynamics in Six Dimensional Space-Time

> "There are only two clouds on the horizon of physics, the problem of black body radiation and the Michelson-Morley experiment. They soon would be gone."
>
> Lord Kelvin, Nobel Laureate

Abstract
This paper describes particle dynamics in a 6 dimensional space-time which is the union of a Minkowski 4 dimensional space-time of special relativity with a 2 dimensional Euclidean coordinate system spanned by imaginary time coordinates. In such a world every elementary particle is mirrored by virtual or ghost particles. In fact, we demonstrate that these are antiparticles moving back in time with velocities exceeding the speed of light and are a consequence of quantum fluctuations of the vacuum field.

Introduction
This paper explores the implications of a 6 dimensional space-time on the kinematics of particle motion. Our conjecture is that the spatial and temporal dimensions should be symmetrical. We further assume that 2 of the time coordinates are imaginary to comport with human experience. Some of the topics discussed include the force, momentum and energy equations in such a frame of reference. In particular, we focus on virtual or ghost, particles that could travel at velocities greater than the speed of light. These are the ideal particles to fold into the imaginary components of time. An estimate of their size is also included. This space-time is the union of a 4 dimensional Minkowski coordinate system of special relativity and a 2 dimensional coordinate system spanned by the new time coordinates.

ASSUMPTIONS

- There are three real space coordinates x, y, z.

- There are three scalar time coordinates t_1, t_2, t_3 (one real, two pure imaginary)

    t_1 = traditional physical time

    $t_2 = i\tau_2$

    $t_3 = i\tau_3$

    where τ_2 and τ_3 are real.

- Each point in space-time is represented by six coordinates x, y, z, t_1, τ_2, τ_3
- Minkowski coordinate system = $Oxyzt_1$; time subspace = $O\tau_2\tau_3$ (Euclidean)
- The elementary particles of physics → $Oxyzt_1$ (4 dimensions)
- Virtual or ghost particles → $Oxyz\tau_2\tau_3$ (5 dimensions)

1. TRANSFORMATION EQUATIONS

The transformation relations of special relativity are based on the invariance of the quantity $x^2 + y^2 + z^2 - c^2 t_1^2$ under Lorentz transformation with respect to coordinate systems S, S', in uniform relative motion.

i.e. $x^2 + y^2 + z^2 - c^2 t_1^2 = x'^2 + y'^2 + z'^2 - c^2 t_1'^2$ where c is the speed of light

If this is modified to

$$x^2 + y^2 + z^2 - c^2 (t_1^2 + t_2^2 + t_3^2) = x'^2 + y'^2 + z'^2 - c^2 (t_1'^2 + t_2'^2 + t_3'^2) \qquad (1)$$

then

$$x^2 + y^2 + z^2 - c^2 (t_1^2 - \tau_2^2 - \tau_3^2) = x'^2 + y'^2 + z'^2 - c^2 (t_1'^2 - \tau_2'^2 - \tau_3'^2) \qquad (2)$$

For consistency with special relativity

$$\tau_2^2 + \tau_3^2 = \tau_2'^2 + \tau_3'^2 \qquad (3)$$

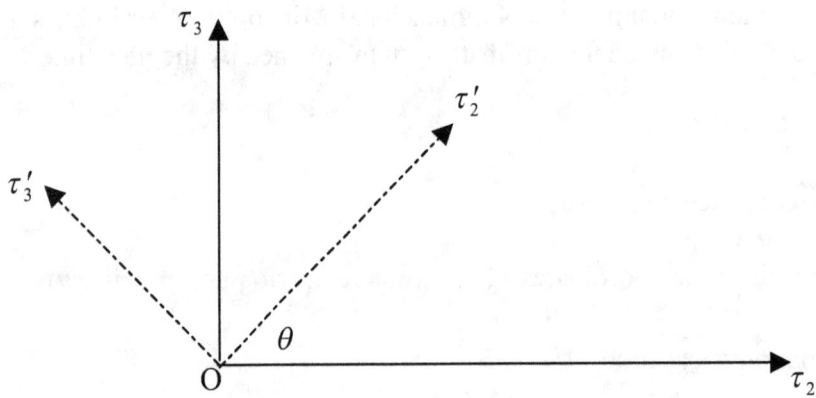

The transformation equations from the primed to the unprimed coordinate system are given by

$$\tau_2' = \cos\theta \tau_2 + \sin\theta \tau_3 \qquad (4a)$$
$$\tau_3' = -\sin\theta \tau_2 + \cos\theta \tau_3 \qquad (4b)$$

where θ is the angle between the $O\tau_2, O\tau_2'$ axes, and corresponds to an orthogonal rotation in 2 dimensions.

The metric of this 6 dimensional world is given by

$$ds^2 = c^2 dt^2 - dx^2 - dy^2 - dz^2 - c^2(d\tau_2^2 + d\tau_3^2) \qquad (5)$$

Chapter 3   Particle Dynamics in Six Dimensional Space-Time

> " There are only two clouds on the horizon of physics, the problem of black body radiation and the Michelson-Morley experiment. They soon would be gone."
>
> Lord Kelvin, Nobel Laureate

Abstract

This paper describes particle dynamics in a 6 dimensional space-time which is the union of a Minkowski 4 dimensional space-time of special relativity with a 2 dimensional Euclidean coordinate system spanned by imaginary time coordinates. In such a world every elementary particle is mirrored by virtual or ghost particles. In fact, we demonstrate that these are antiparticles moving back in time with velocities exceeding the speed of light and are a consequence of quantum fluctuations of the vacuum field.

Introduction

This paper explores the implications of a 6 dimensional space-time on the kinematics of particle motion. Our conjecture is that the spatial and temporal dimensions should be symmetrical. We further assume that 2 of the time coordinates are imaginary to comport with human experience. Some of the topics discussed include the force, momentum and energy equations in such a frame of reference. In particular, we focus on virtual or ghost, particles that could travel at velocities greater than the speed of light. These are the ideal particles to fold into the imaginary components of time. An estimate of their size is also included. This space-time is the union of a 4 dimensional Minkowski coordinate system of special relativity and a 2 dimensional coordinate system spanned by the new time coordinates.

ASSUMPTIONS
- There are three real space coordinates x, y, z.

- There are three scalar time coordinates t_1, t_2, t_3 (one real, two pure imaginary)

    t_1 = traditional physical time

    $t_2 = i\,\tau_2$

    $t_3 = i\,\tau_3$

    where τ_2 and τ_3 are real.

- Each point in space-time is represented by six coordinates x, y, z, t_1, τ_2, τ_3
- Minkowski coordinate system = $Oxyzt_1$; time subspace = $O\tau_2\tau_3$ (Euclidean)
- The elementary particles of physics → $Oxyzt_1$ (4 dimensions)
- Virtual or ghost particles → $Oxyz\tau_2\tau_3$ (5 dimensions)

1. TRANSFORMATION EQUATIONS

The transformation relations of special relativity are based on the invariance of the quantity $x^2 + y^2 + z^2 - c^2 t_1^2$ under Lorentz transformation with respect to coordinate systems S, S', in uniform relative motion.

i.e. $\quad x^2 + y^2 + z^2 - c^2 t_1^2 = x'^2 + y'^2 + z'^2 - c^2 t_1'^2 \quad$ where c is the speed of light

If this is modified to

$$x^2 + y^2 + z^2 - c^2(t_1^2 + t_2^2 + t_3^2) = x'^2 + y'^2 + z'^2 - c^2(t_1'^2 + t_2'^2 + t_3'^2) \tag{1}$$

then

$$x^2 + y^2 + z^2 - c^2(t_1^2 - \tau_2^2 - \tau_3^2) = x'^2 + y'^2 + z'^2 - c^2(t_1'^2 - \tau_2'^2 - \tau_3'^2) \tag{2}$$

For consistency with special relativity

$$\tau_2^2 + \tau_3^2 = \tau_2'^2 + \tau_3'^2 \tag{3}$$

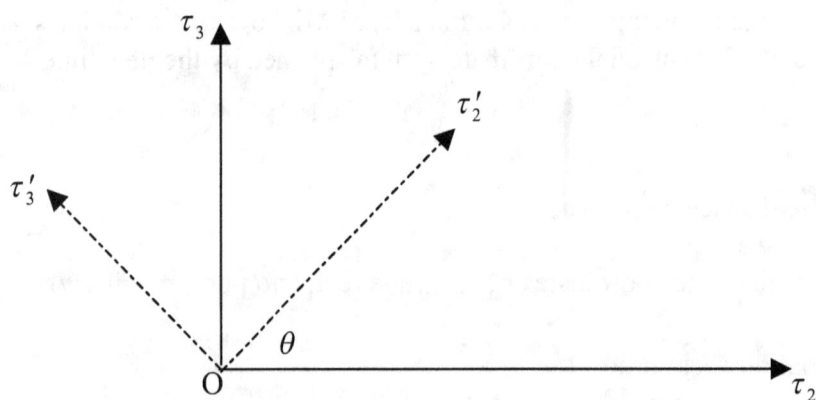

The transformation equations from the primed to the unprimed coordinate system are given by

$$\tau_2' = \cos\theta\,\tau_2 + \sin\theta\,\tau_3 \tag{4a}$$
$$\tau_3' = -\sin\theta\,\tau_2 + \cos\theta\,\tau_3 \tag{4b}$$

where θ is the angle between the $O\tau_2, O\tau_2'$ axes, and corresponds to an orthogonal rotation in 2 dimensions.

The metric of this 6 dimensional world is given by

$$ds^2 = c^2 dt^2 - dx^2 - dy^2 - dz^2 - c^2(d\tau_2^2 + d\tau_3^2) \tag{5}$$

2. MOMENTUM EQUATIONS

The relativistic momentum of a particle with rest mass m_o and velocity V is given by

$$p = \frac{m_o V}{\sqrt{1 - \frac{V^2}{c^2}}} \qquad (6)$$

for time given as t_1, t_2, t_3 this can be rewritten as

$$p(t_1) = \frac{m_o V_1}{\sqrt{1 - \frac{V_1^2}{c^2}}} \qquad (6a)$$

$$p(t_2) = \frac{m_o V_2}{\sqrt{1 - \frac{V_2^2}{c^2}}} \qquad (6b)$$

$$p(t_3) = \frac{m_o V_3}{\sqrt{1 - \frac{V_3^2}{c^2}}} \qquad (6c)$$

where V_1, V_2, V_3 are given as follows for a real distance r

$$V_1 = \frac{dr}{dt_1}$$

$$V_2 = \frac{dr}{dt_2} = \frac{dr}{i d\tau_2}$$

$$V_3 = \frac{dr}{dt_3} = \frac{dr}{i d\tau_3}$$

By defining real velocities u_2 and u_3 as follows

$$u_2 = \frac{dr}{d\tau_2} \qquad (7a)$$

$$u_3 = \frac{dr}{d\tau_3} \tag{7b}$$

then V_2 and V_3 can be written as

$$V_2 = \frac{1}{i} u_2 \tag{7c}$$

$$V_3 = \frac{1}{i} u_3 \tag{7d}$$

and thus the momentum equations can be written as

$$p(t_1) = \frac{m_o V_1}{\sqrt{1 - \frac{V_1^2}{c^2}}} \tag{8a}$$

$$p(t_2) = \frac{1}{i} \frac{m_o u_2}{\sqrt{1 + \frac{u_2^2}{c^2}}} \tag{8b}$$

$$p(t_3) = \frac{1}{i} \frac{m_o u_3}{\sqrt{1 + \frac{u_3^2}{c^2}}} \tag{8c}$$

Let us write,

$$\begin{aligned} p(t_2) &= i\,\vartheta_2(\tau_2) \\ p(t_3) &= i\,\vartheta_3(\tau_3) \end{aligned} \tag{9}$$

where $\vartheta(\tau_2), \vartheta(\tau_3)$ are real and

$$\vartheta_2(\tau_2) = -\frac{m_0 u_2}{\sqrt{1 + \frac{u_2^2}{c^2}}} \tag{9a}$$

$$\vartheta_3(\tau_3) = -\frac{m_0 u_3}{\sqrt{1 + \frac{u_3^2}{c^2}}} \tag{9b}$$

Therefore in the real (τ_2, τ_3) coordinate system, the particle has (real) momentum components $\vartheta_2(\tau_2), \vartheta_3(\tau_3)$. These latter expressions correspond to tachyons, particles moving at velocities greater than the speed of light. The minus signs mean the particles are moving backward in τ – time. This is the signature feature of antiparticles.

3. PARTICLE TRAJECTORY

Traditional physics: $r = r(t_1)$

New physics: $r = r(t_1, \tau_2, \tau_3)$
where

$$\frac{\partial r}{\partial \tau_2} = u_2 \qquad (10a)$$

$$\frac{\partial r}{\partial \tau_3} = u_3 \qquad (10b)$$

and the following constrains are applied

τ_2
(see Section 4)

u_2 = constant

u_3 = constant

This constitutes an orthogonal grid of lines parallel to the τ_2, τ_3 time axes as depicted above.
Then

$$r = u_2 \tau_2 + u_3 \tau_3 + \psi(t_1) \qquad (11)$$

where $\psi(t_1)$ is the observable trajectory in the $Oxyz$ coordinate system.

4. FORCE EQUATIONS

The relativistic equation of motion for a particle of rest mass m_o moving with velocity V is given by

$$F = \frac{d}{dt}\left(\frac{m_o V}{\sqrt{1-\frac{V^2}{c^2}}}\right) \qquad (12)$$

This is replaced by three equations corresponding to times t_1, t_2, t_3

$$F(t_1) = \frac{d}{dt_1}\left(\frac{m_o V_1}{\sqrt{1-\frac{V_1^2}{c^2}}}\right) \quad (12a)$$

$$F(t_2) = \frac{d}{dt_2}\left(\frac{m_o V_2}{\sqrt{1-\frac{V_2^2}{c^2}}}\right) \quad (12b)$$

$$F(t_3) = \frac{d}{dt_3}\left(\frac{m_o V_3}{\sqrt{1-\frac{V_3^2}{c^2}}}\right) \quad (12c)$$

From the definition of τ_2, τ_3 and u_2, u_3

$$F(t_2) = F(i\tau_2) = \frac{d}{i\,d\tau_2}\left(\frac{m_o V_2}{\sqrt{1-\frac{V_2^2}{c^2}}}\right) = \frac{d}{i\,d\tau_2}\left(\frac{m_o u_2}{i\sqrt{1+\frac{u_2^2}{c^2}}}\right) = -\frac{d}{d\tau_2}\left(\frac{m_o u_2}{\sqrt{1+\frac{u_2^2}{c^2}}}\right)$$

$$= \frac{d\vartheta_2(\tau_2)}{d\tau_2} \quad (12d)$$

$$F(t_3) = F(i\tau_3) = \frac{d}{i\,d\tau_3}\left(\frac{m_o V_3}{\sqrt{1-\frac{V_3^2}{c^2}}}\right) = \frac{d}{i\,d\tau_3}\left(\frac{m_o u_3}{i\sqrt{1+\frac{u_3^2}{c^2}}}\right) = -\frac{d}{d\tau_3}\left(\frac{m_o u_3}{\sqrt{1+\frac{u_3^2}{c^2}}}\right)$$

$$= \frac{d\vartheta_3(\tau_3)}{d\tau_3} \quad (12e)$$

In summary, the total momentum is

$$p = p(t_1) + p(t_2) + p(t_3) \tag{13}$$

where

$$p(t_1) = \text{real}$$

$$p(t_2) = \text{pure imaginary}$$

$$p(t_3) = \text{pure imaginary}$$

and the total force is

$$F = F(t_1) + F(t_2) + F(t_3) \tag{14}$$

where

$$F(t_1) = \text{real}$$

$$F(t_2) = \text{real}$$

$$F(t_3) = \text{real}$$

Since there are no observable forces in the (τ_2, τ_3) plane, $F(t_2) = F(t_3) = 0$

$$\frac{d}{d\tau_2} \vartheta_2(\tau_2) = \frac{d}{d\tau_3} \vartheta_3(\tau_3) = 0 \tag{15}$$

$$\therefore \quad \vartheta_2(\tau_2) = \text{constant}$$
$$\vartheta_3(\tau_3) = \text{constant}$$

$$\therefore \quad u_2 = \text{constant}$$
$$u_3 = \text{constant}$$

5. ENERGY EQUATIONS

$$E = m c^2 \tag{16}$$

where

$$m = \frac{m_o}{\sqrt{1-\frac{v^2}{c^2}}} \quad (16a)$$

combining (16, 16a) yields

$$E = \frac{m_o c^2}{\sqrt{1-\frac{v^2}{c^2}}} \quad (17)$$

also

$$p = mv$$

$$\therefore p = \frac{m_o v}{\sqrt{1-\frac{v^2}{c^2}}} \quad (18)$$

Taking the ratio of E and p using equations (17, 18) and re-arranging terms yields

$$\frac{E}{p} = \frac{c^2}{v}$$

or

$$v = \frac{c^2 p}{E} \quad (19)$$

substituting (19) into the equation (17) for E above yields

$$E = \frac{m_o c^2}{\sqrt{1-\frac{c^4 p^2}{c^2 E^2}}} = \frac{m_o c^2}{\sqrt{1-\frac{c^2 p^2}{E^2}}}$$

squaring and re-arranging

$$E^2 = \frac{m_o^2 c^4}{1-\frac{c^2 p^2}{E^2}}$$

$$E^2 \left(1-\frac{c^2 p^2}{E^2}\right) = m_o^2 c^4$$

$$E^2 - c^2 p^2 = m_o^2 c^4$$

$$E^2 = m_o^2 c^4 + c^2 p^2 = c^2 \left(m_o^2 c^2 + p^2 \right)$$

then taking square root of both sides yields

$$E = \pm c \sqrt{m_o^2 c^2 + p^2} \tag{20}$$

This can be written in terms of t_1, t_2, and t_3 as follows

$$E = \pm c \sqrt{m_o^2 c^2 + p(t_1)^2} \pm c \sqrt{m_o^2 c^2 + p(t_2)^2} \mp c \sqrt{m_o^2 c^2 + p(t_3)^2}$$

Since m_o is real, then $p(t_1)$ is real but $p(t_2)$ and $p(t_3)$ are imaginary.

Note that if $p(t_1)$, $p(t_2)$, $p(t_3)$ are zero, then this equation reduces to

$$E = \pm m_o c^2 \pm m_o c^2 \mp m_o c^2 = \pm m_o c^2 \tag{20a}$$

which is the usual expression for rest energy of a particle.

6. PARTICLE WAVE FUNCTION

Particle Wave function : $\Psi(x,y,z,t_1,\tau_2,\tau_3) = \psi(x,y,z,t_1)\phi(\tau_1,\tau_2)$ (21)
where,

$\psi = \psi(x,y,z,t_1)$ = particle wave function in Minkowski space
$\phi = \phi(\tau_1,\tau_2) = \quad$ wave function of ghost particle

and,

$$i\hbar \frac{\partial \phi}{\partial t_2} = E_2 \phi$$

$$i\hbar \frac{\partial \phi}{\partial t_3} = E_3 \phi \tag{22}$$

where E_2, E_3 are real and,

$$E_2 = \pm c\sqrt{m_0^2 c^2 + p^2(t_2)} = \pm c\sqrt{m_0^2 c^2 - \vartheta_2^2(\tau_2)}$$
$$E_3 = \pm c\sqrt{m_0^2 c^2 + p^2(t_3)} = \pm c\sqrt{m_0^2 c^2 - \vartheta_3^2(\tau_3)} \qquad (23)$$

The above expressions can be further simplified using the explicit expressions (9a, 9b) for $\vartheta_2(\tau_2), \vartheta_3(\tau_3)$

$$E_2 = \pm \frac{m_0 c^2}{\sqrt{1 + \frac{u_2^2}{c^2}}} \qquad (23a)$$

$$E_3 = \pm \frac{m_0 c^2}{\sqrt{1 + \frac{u_3^2}{c^2}}} \qquad (23b)$$

Using a separation of variables technique,

$$\phi = \exp(\frac{E_2 \tau_2 + E_3 \tau_3}{\hbar}) \qquad (24)$$

$$\Psi = \psi \exp(\frac{E_2 \tau_2 + E_3 \tau_3}{\hbar}) \qquad (25)$$

$\therefore \Psi$ is bounded as $t \to \infty$ if $E_2, E_3 < 0$.

7. UNCERTAINTY PRINCIPLE

$$\therefore \int |\Psi|^2 \, dV_6 = \int |\psi|^2 \, dV_4 \int_0^\infty \int_0^\infty \exp 2(\frac{E_2 \tau_2 + E_3 \tau_3}{\hbar}) \, d\tau_2 d\tau_3$$

$$= \frac{\hbar^2}{4 E_2 E_3} \int |\psi|^2 \, dV_4 = \frac{\hbar^2}{4 E_2 E_3} \qquad (26)$$

Since, the particle wave function ψ in 4 dimensions must satisfy the normalization condition,

$$\int |\psi|^2 \, dV_4 = 1 \qquad (27)$$

For spin 1/2 particles ψ satisfies the Dirac equation, while for spin 0 particles ψ must satisfy the Klein-Gordon equation. Therefore for large $|E_2|, |E_3|$ the probability of finding the ghost particles decrease.
But,

$$|E_2|, |E_3| < m_0 c^2 \tag{28}$$

$$\therefore \int |\Psi|^2 \, dV_6 > \frac{\hbar^2}{4 m_0^2 c^4} \tag{29}$$

$\therefore$ The uncertainty in the energy of these particles is

$$\Delta E_{2 \leftrightarrow 3} \approx 2 m_0 c^2$$

The uncertainty in the corresponding time interval is given by,

$$\Delta \tau_{2 \leftrightarrow 3} \approx \frac{\hbar}{2 m_0 c^2} \tag{30}$$

Therefore the size of these particles is

$$r \approx c \, \Delta \tau_{2 \leftrightarrow 3} = \frac{\hbar}{2 m_0 c} \tag{31}$$

8. TIME REVERSAL

The 2 dimensional subspace (τ_2, τ_3) is Euclidean, and the motion of a particle of mass m and charge e in an electromagnetic field **E**, **H** with velocity **v** is given by,

$$m \frac{d}{dt} \mathbf{v} = e \left(\mathbf{E} + \frac{1}{c} \mathbf{v} \wedge \mathbf{H} \right)$$

Similarly, a particle of charge $-e$ will be governed by the equation,

$$m \frac{d}{dt} \mathbf{v} = -e \left(\mathbf{E} + \frac{1}{c} \mathbf{v} \wedge \mathbf{H} \right) \tag{32}$$

Put, $t' = -t$

$$\therefore \mathbf{r}' = -\mathbf{r}$$

and, $\mathbf{v}' = \dfrac{d}{dt'} \mathbf{r}' = \dfrac{d}{d(-t)}(-\mathbf{r}) = \dfrac{d}{dt} \mathbf{r} = \mathbf{v}$

Then,

$$m\frac{d}{dt'}\mathbf{v} = e(\mathbf{E} + \frac{1}{c}\mathbf{v}\wedge\mathbf{H})\tag{33}$$

which is equivalent to a particle of charge e moving back in time.

9. DISCUSSION

1. ***The significance of the imaginary times*** t_2, t_3. The virtual particles are not subject to any restrictions on their velocities as they populate the imaginary time subspace (τ_2, τ_3).

2. ***The covariance of the equations*** is assured because the space (x, y, z, t_1) is Lorentzian and the the elementary particles populate this space. The (τ_2, τ_3) space is Euclidean. The virtual particles inhabit the latter space and are not affected by gravity.

3. ***The energy expressions*** E_2, E_3 are real and give an estimate for the size of the virtual particles, which is capable of being measured in a bubble chamber.

4. ***The initial purpose of this paper*** was to demonstrate a possible application of the math model of 6D space-time to enable investigation of the virtual particles as a detectable entity.

APPENDIX

- **Elastic Collisions**

Conservation of Momentum and Energy in ***independent*** real (t_1, τ_2, τ_3) time domains

VELOCITIES BEFORE IMPACT	$(u_1; u_2, u_3)$	$(v_1; v_2, v_3)$
	$m_0 \circ \rightarrow$	$M_0 \circ \rightarrow$
VELOCITIES AFTER IMPACT	$(u_1'; u_2', u_3')$	$(v_1'; v_2', v_3')$

Momentum equations, Section 2:

$$p(t_1; u_1) + p(t_1; v_1) = p(t_1; u_1') + p(t_1; v_1')$$
$$\vartheta_2(\tau_2; u_2) + \vartheta_2(\tau_2; v_2) = \vartheta_2(\tau_2; u_2') + \vartheta_2(\tau_2; v_2')$$
$$\vartheta_3(\tau_3; u_3) + \vartheta_3(\tau_3; v_3) = \vartheta_3(\tau_3; u_3') + \vartheta_3(\tau_3; v_3')$$

Energy equations, Section 6:

$$E(t_1;u_1) + E(t_1;v_1) = E(t_1;u_1') + E(t_1;v_1')$$
$$E_2(\tau_2;u_2) + E_2(\tau_2;v_2) = E_2(\tau_2;u_2') + E_2(\tau_2;v_2')$$
$$E_3(\tau_3;u_3) + E_3(\tau_3;v_3) = E_3(\tau_3;u_3') + E_3(\tau_3;v_3')$$

These correspond to the following equations for the six unknowns : $(u_1';u_2',u_3')$, $(v_1';v_2',v_3')$

$$\frac{m_0 u_1}{\sqrt{1-\frac{u_1^2}{c^2}}} + \frac{M_0 v_1}{\sqrt{1-\frac{v_1^2}{c^2}}} = \frac{m_0 u_1'}{\sqrt{1-\frac{u_1'^2}{c^2}}} + \frac{M_0 v_1'}{\sqrt{1-\frac{v_1'^2}{c^2}}} \qquad (34)$$

$$\frac{m_0 u_2}{\sqrt{1+\frac{u_2^2}{c^2}}} + \frac{M_0 v_2}{\sqrt{1+\frac{v_2^2}{c^2}}} = \frac{m_0 u_2'}{\sqrt{1+\frac{u_2'^2}{c^2}}} + \frac{M_0 v_2'}{\sqrt{1+\frac{v_2'^2}{c^2}}} \qquad (35)$$

$$\frac{m_0 u_3}{\sqrt{1+\frac{u_3^2}{c^2}}} + \frac{M_0 v_3}{\sqrt{1+\frac{v_3^2}{c^2}}} = \frac{m_0 u_3'}{\sqrt{1+\frac{u_3'^2}{c^2}}} + \frac{M_0 v_3'}{\sqrt{1+\frac{v_3'^2}{c^2}}} \qquad (36)$$

$$\pm\frac{m_0 c^2}{\sqrt{1-\frac{u_1^2}{c^2}}} \pm \frac{M_0 c^2}{\sqrt{1-\frac{v_1^2}{c^2}}} = \pm\frac{m_0 c^2}{\sqrt{1-\frac{u_1'^2}{c^2}}} \pm \frac{M_0 c^2}{\sqrt{1-\frac{v_1'^2}{c^2}}} \qquad (37)$$

$$\pm\frac{m_0 c^2}{\sqrt{1+\frac{u_2^2}{c^2}}} \pm \frac{M_0 c^2}{\sqrt{1+\frac{v_2^2}{c^2}}} = \pm\frac{m_0 c^2}{\sqrt{1+\frac{u_2'^2}{c^2}}} \pm \frac{M_0 c^2}{\sqrt{1+\frac{v_2'^2}{c^2}}} \qquad (38)$$

$$\pm \frac{m_0 c^2}{\sqrt{1+\frac{u_3^2}{c^2}}} \pm \frac{M_0 c^2}{\sqrt{1+\frac{v_3^2}{c^2}}} = \pm \frac{m_0 c^2}{\sqrt{1+\frac{u_3'^2}{c^2}}} \pm \frac{M_0 c^2}{\sqrt{1+\frac{v_3'^2}{c^2}}} \qquad (39)$$

For a given $(u_1; u_2, u_3)$, $(v_1; v_2, v_3)$ these equations can be solved numerically using for example a least squares approximation. However, $u_2, u_3; v_2, v_3$ are unknown and in accordance with Section 3, equations (1-6) must be solved as "contour" solutions for specified constant values of $u_2, u_3; v_2, v_3$ all greater than the velocity of light.

Chapter 4 Ordered Space-Time Derivatives in Cosmology and Particle Physics

" Those who are not shocked when they first come across quantum theory cannot possibly have understood it."

Werner Heisenberg, Nobel Laureate

Abstract :

This paper describes a Maclaurin power series representation for the fundamental forces of nature immediately after the big bang. We use an infinite sequence of space and time derivatives of increasing order to describe physical phenomena from subatomic particles to spiral nebulae. The derivatives can be conceptualized either as an infinite dimensional vector or a scalar function of space and time in the form of a Maclaurin series, the coefficients of which correspond to the higher order derivatives. We believe our presentation is a more robust version of Dirac's large number hypothesis aimed at a deeper understanding of the nexus between general relativity and atomic physics.

We estimate the temperature of the universe at the time of the big-bang, and the electromagnetic energy released during the explosion. Dirac's model requires an ever expanding universe, with contractions precluded. We report similar conclusions, where in future epochs some of the space time derivatives are positive definite.

Introduction

In the immediate aftermath of the big-bang, space-time was ripped asunder and space and time came into being. The latent energy of the explosion manifested in the form of a unified electrogravity field became decoupled into electromagnetic radiation and matter. This act of creation was followed by the splitting of the electroweak force, making possible the formation of deuterium and the heavy nucleii, needed for the assembling of atoms and molecules. These processes gave structure to the early universe.

These events provided the seeds for the future evolution of the observable universe. In the few moments of creation, the basic laws of physics as presently understood came into existence: They are based on a few fundamental "constants" of nature → such as G the universal constant of gravitation, c the velocity of light, Planck's constant $\hbar$, the charge e and rest mass m_e of the electron, and proton mass m_p. In turn these constants of nature, provide other physical constants such as the fine structure constant α, Boltzmann's constant k, and Avagadro's number N required for a more complete understanding of our universe.

Using this information, we generate an infinite sequence of space-time derivatives of increasing order, to model the basic forces of nature, as summarized in the flowchart below.

Specifically, the derivatives are with respect to space variables, (x, y, z) and time t. For purposes of simplicity, we restrict our attention to spherically symmetric universes characterized by a single spatial variable and demonstrate in some detail the consequences of our postulates restricting attention to the first four space-time derivatives, corresponding to the 4 fundamental forces of nature:

- **Strong force**, that holds an atomic nucleus in a stable configuration against the force of repulsion between the protons in the nucleus. It was mediated in the early theory by pi-mesons whose properties are described by short range exchange forces of the type studied by Yukawa. It has a range of 10^{-13} cm. This is the macro approach we adopt here, rather than using the more detailed picture of the gluon as the exchange particle mediating the the color force between the quarks that make up the protons and neutrons in the nucleus. The strong force is the strongest of the 4 fundamental forces.

- **Electromagnetic force**, described by Maxwell's equations has an infinite range and in quantum electrodynamics is mediated by exchange forces involving photons. The electromagnetic force holds atoms and molecules together. In fact, the forces of attraction and repulsion of electric charges are so dominant over the other three fundamental forces that they can be considered to be negligible in determining the atomic and molecular structure of matter. Even magnetic effects are usually apparent only at high resolutions, and as small corrections. The electromagnetic force is about 1/137 the strength of the strong force.

- **Weak force**, involves the exchange of intermediate vector bosons and has a range of about 10^{-16} cm, which is about 1/1000 of the diameter of a proton. It is about 10^{-5} times as strong as the strong force, but is of fundamental importance to an understanding of the structure of the early universe, in the formation of deuterium, the buildup of heavy nucleii and in beta decay. The electromagnetic force and the weak nuclear force are 2 different aspects of a single unified electroweak force

- **Gravitational force** is the weakest of the 4 forces The gravitational force is a force of infinite range which obeys Newton's inverse square law of gravitation, and is of the same form as the electromagnetic force. It is about 10^{-39} times the strong force.

1. Large Number Hypothesis

In 1937, Dirac proposed his large numbers hypothesis to bridge modern cosmology and particle physics, using the properties of certain dimensionless numbers. An example of a dimensionless number is, the ratio of the rest mass of a proton to the rest mass of an electron ≈ 1833. Another dimensionless number in spectroscopy that connects Planck's constant to the electronic charge, the fine structure constant ≈ 1/137 which is independent of the units used. A particularly important dimensionless number is the ratio of the electric force to the gravitational force of attraction between an electron and a proton. Since in either case, the force is inversely proportional to the square of the distance, (Coulomb's law of electrostatics & Newton's law of gravitation), the ratio of these 2 forces is independent of the distance. The ratio is an extremely large number ≈ 10^{39}, which is again independent of the system of units used. Physicists have long pondered as to what significance can be attached to such a large number. This number is also approximately the age of the universe when expressed in atomic units of time, the present age being about 18 billion years according to Hubble's Law.

He argued that this coincidence could be explained by Newton's constant of gravitation decreasing as the universe aged. . Such a variation lies outside general relativity, in which G must be constant but can be incorporated by a fairly simple modification of the theory. Other models, including the Brans-Dicke theory of gravity and some versions of super string theory also predict physical "constants" that vary with time.

The number of fundamental particles - electrons, protons, and neutrons - in the universe is about 10^{78}, the square of the age of the universe. The number of particles in the universe will be increasing proportionally to the square of the age of the universe. Thus new matter must be continually created, similar to the requirements of the quasi-steady state theory of Hoyle and Narlikar. Dirac requires all large numbers to be related to the age of the universe and must increase with age. This leads to a universe that is always expanding and never contracting, thus precluding singularities in future..

Our approach to synthesizing gravitation and atomic physics, is to use a dimensionless **function** of space and time, instead of a pure number. We also depart from Dirac in using **derivatives** → first, second, third, fourth,…, derivatives of space and time rather than exponents of a number, in this case the age of the universe.

We deviate from Dirac's model in

- Using space-time derivatives of dimensionless **functions** as our fundamental point of departure, rather than **pure** dimensionless **numbers.**
- First derivatives, recreate the fundamental relations of special relativity, with particular reference to the energy and momentum of a particle .
- The second derivatives provide a forum for studying planetary motion both classically and quantum mechanically.
- We use a function Ω analogous to the age of the universe T in Dirac's analysis. The function Ω is a measure of the deviation from flat space-time due to gravity. Its third derivatives with respect to space and time leads to a model closely analogous to the equations of fluid dynamics, giving rise to shock waves.
- We believe our approach is more robust than the model proposed by Dirac and enables us to validate the strong force in atomic physics which is about 100 times stronger than the electromagnetic forces.
- We provide an estimate for the temperatures of the early universe and its electromagnetic energy, based on a Maclaurin series expansion of Ω

2. First Derivatives

Using derivatives of space-time functions rather than a dimensionless number, gives a more dynamic picture of the universe. The simplest building blocks of our model are the energy and momentum of a particle. If the particle is a photon, then its energy is directly proportional to the momentum. The energy ~ momentum curve is a pair of straight lines, through the origin with slopes numerically equal to the speed of light. For particles with nonzero rest mass, these represent the asymptotes for the energy ~ momentum curves. The energy intercepts are a measure of the rest energy or rest mass of the particle.

On the other hand, a plot of $(energy)^2$ ~ $(momentum)^2$ for a single photon yields a straight line through the origin with slope c^2, while a particle will correspond to a parallel straight line with a vertical intercept of $(rest\ energy)^2$.

The wave velocity w of a wave packet associated with a particle of mass m moving with a velocity v is given by:

$$w = \frac{c^2}{v} \quad \Rightarrow \quad wv = c^2 = f_1(c) \tag{1}$$

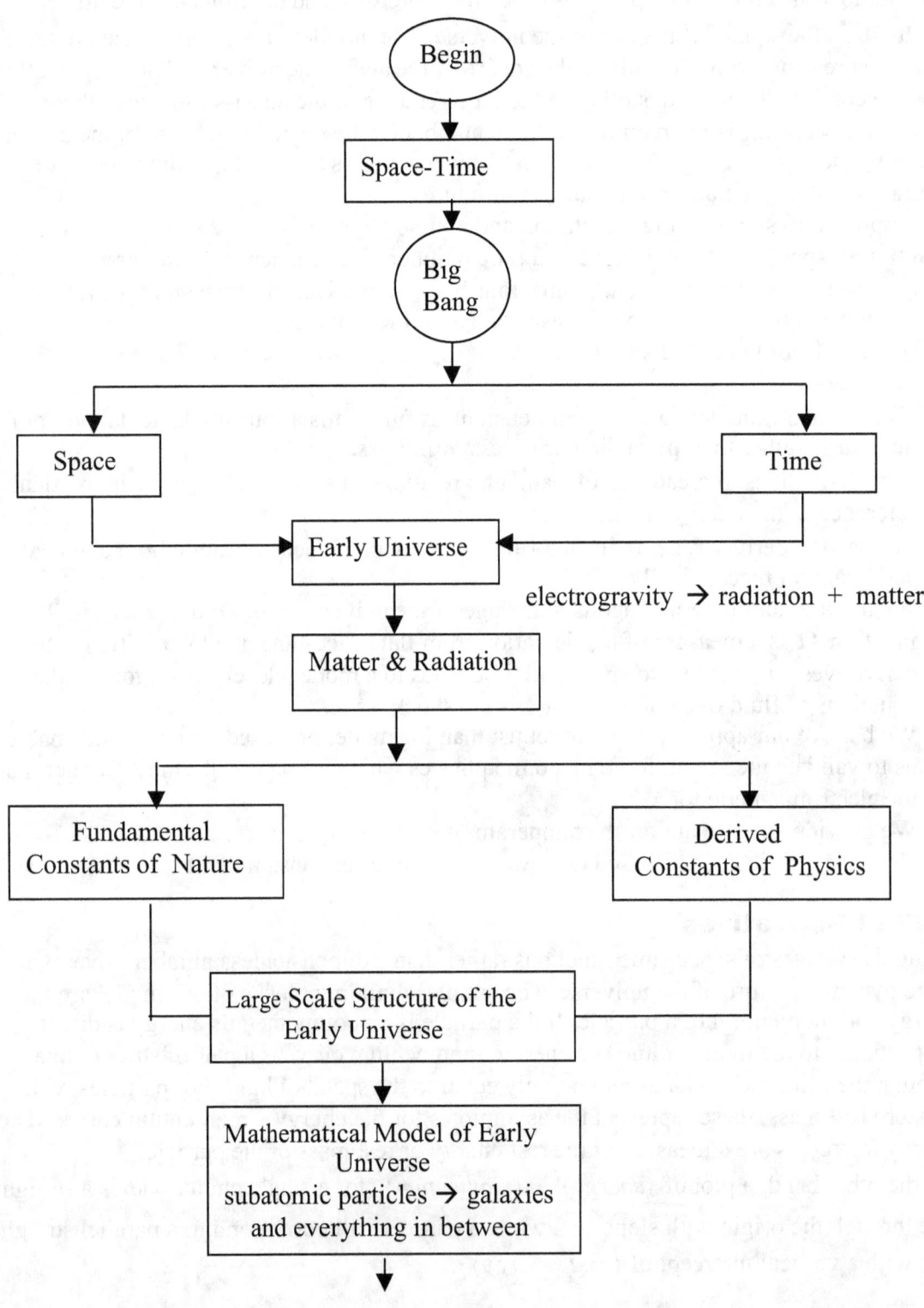

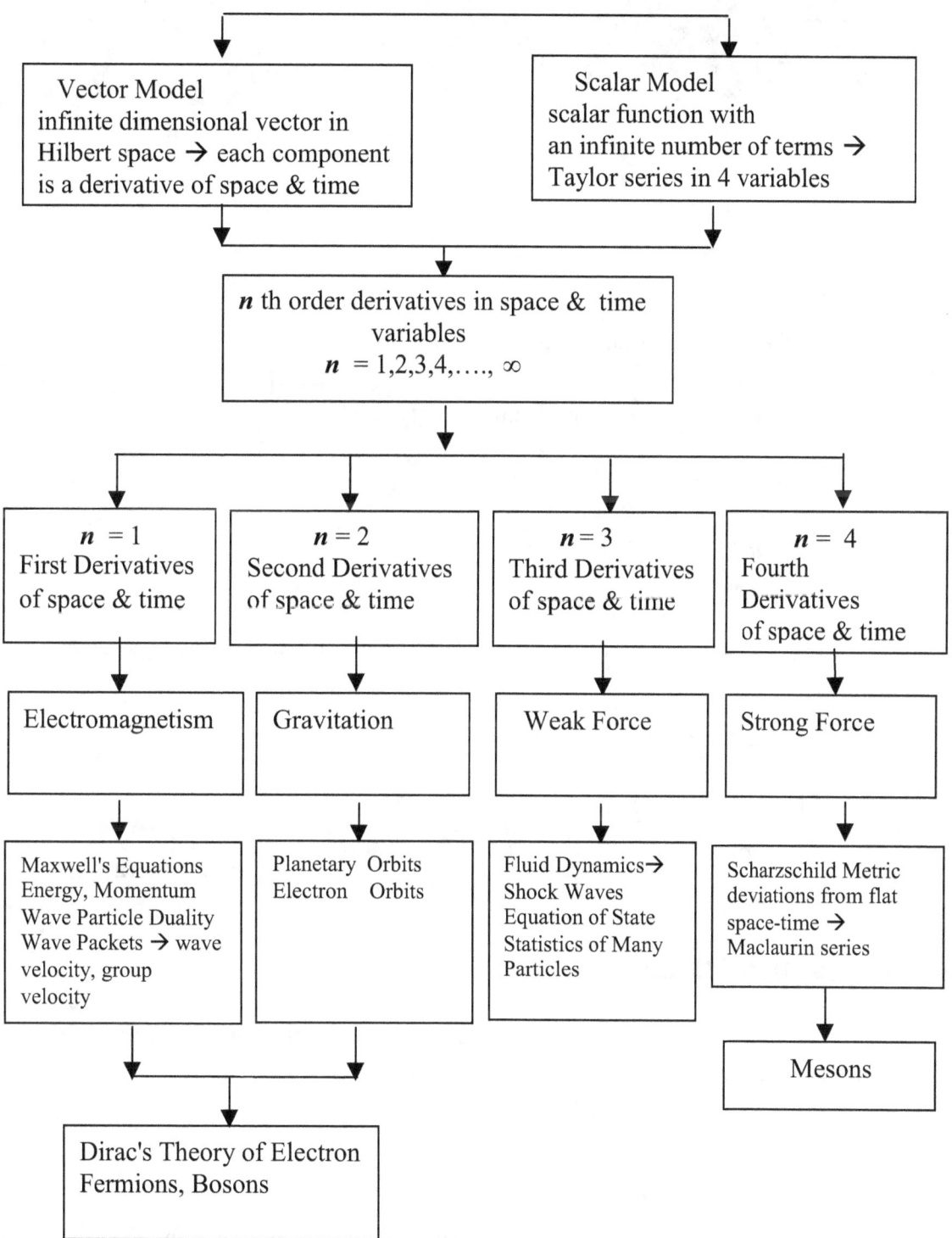

- The relation between the frequency ν and the wavelength λ of electromagnetic waves in a vacuum is

$$c = \nu \lambda_0 \quad \Rightarrow \quad \lambda_0 = \frac{c}{\nu}$$

$$\therefore \quad \frac{d\lambda_0}{d(\frac{1}{\nu})} = c$$

i.e. $\quad \dfrac{d\lambda_0}{dT_0} = c \qquad (2)$

where T is the period of oscillations: $T_0 = 1/\nu$

- The energy E of a photon of momentum p is given by

$$E = cp$$

$$\therefore \quad \frac{dE}{dp} = c \qquad (3)$$

The energy E of an elementary particle of rest mass m_0, having momentum p is

$$E = \pm c\sqrt{m_0^2 c^2 + p^2}$$

$$\therefore \quad E^2 = c^2(m_0^2 c^2 + p^2)$$

$$\therefore \quad \frac{dE^2}{dp^2} = c^2 \qquad (4)$$

3. Second Derivatives

- Planetary motion: Consider a particle of mass m, executing a circular orbit of radius r around a massive star of mass M. Then its period T_1 is given by

$$T_1 = \frac{2\pi r}{v}$$

where v is the orbital speed of the particle. The equation of motion along the radial direction becomes,

$$\frac{mv^2}{r} = \frac{GmM}{r^2} \quad \Rightarrow \quad v^2 = \frac{GM}{r}$$

where the gravitational force of attraction is balanced by the outward pull of the centrifugal force

Eliminating v,

$$r^3 = \frac{GM}{4\pi^2} T_1^2$$

Differentiating twice we have,

$$\frac{d^2 r^3}{dT_1^2} = \frac{GM}{2\pi^2} \qquad (5)$$

The de Broglie wavelength associated with the particle is given by

$$\lambda_1 = \frac{h}{mv} = \frac{\hbar T_1}{mr} = \frac{\hbar}{m}\left(\frac{4\pi^2}{GM}\right)^{1/3} T_1^{1/3} \qquad (6)$$

$$\frac{d\lambda_1}{dT_1} = \frac{\hbar}{3m}\left(\frac{4\pi^2}{GM}\right)^{1/3} T_1^{-\frac{2}{3}} \qquad (7)$$

$$\therefore \quad \frac{d^2\lambda_1}{dT_1^2} = -\frac{2\hbar}{9m}\left(\frac{4\pi^2}{GM}\right)^{1/3} T_1^{-\frac{5}{3}} \qquad (8)$$

Again,

$$v^2 = \frac{GM}{r} = GM\left(\frac{4\pi^2}{GM}\right)^{\frac{1}{3}} T_1^{-\frac{2}{3}} \quad \Rightarrow \quad T_1 = \frac{2\pi GM}{v^3}$$

Substituting,

$$\frac{d\lambda_1}{dT_1} = \frac{\hbar}{3mGM} v^2 = \frac{\hbar c^4}{3mGMw^2} = f_1(c) \qquad (9)$$

$$\frac{d^2\lambda_1}{dT_1^2} = -\frac{\hbar}{9\pi m G^2 M^2} v^5 = -\frac{\hbar c^{10}}{9\pi G^2 M^2 w^5} = f_2(c) \qquad (10)$$

where from (1), $\quad w = \dfrac{c^2}{v}$

is the velocity of propagation of the wave packet associated with the particle of mass m, and is greater than c.

- Bohr model of the hydrogen atom: an electron of charge $-e$ and mass m is describing a circular orbit of radius r about a positively charged nucleus of charge $+e$. Then the energy equation gives,

$$E = \text{Kinetic Energy} + \text{Potential Energy} = \frac{1}{2}mv^2 - \frac{e^2}{r}$$

But, the radial equation of motion is,

$$\frac{mv^2}{r} = \frac{e^2}{r^2} \tag{11}$$

Substituting for v^2,

$$E = \frac{1}{2}(\frac{e^2}{r}) - \frac{e^2}{r} = -\frac{e^2}{2r} \tag{12}$$

Using the Sommerfeld quantization condition,

$$\oint p_\theta \, d\theta = nh \tag{13}$$

where p_θ is the orbital angular momentum, n is an integer and h is Planck's constant. The integration is over the entire circular orbit.

$$\therefore \quad \oint (mvr) \, d\theta = nh \quad \Rightarrow \quad mvr = \frac{nh}{2\pi} \tag{14}$$

A straightforward mathematical manipulation using equations (11) & (14) shows,

$$r = \frac{n^2 h^2}{4\pi^2 me^2} \tag{15}$$

and,

$$E = -\frac{2\pi^2 me^4}{n^2 h^2}, \quad n = 1,2,3,.... \tag{16}$$

This is the energy associated with a stationary orbit characterized by the quantum number n.

According to Bohr, if the electron jumps from an orbit characterized by the quantum number n_1 to an orbit characterized by n_2, then it will emit a photon of frequency v given by

$$E(n_1) - E(n_2) = hv$$

Therefore,

$$v = \frac{2\pi^2 me^4}{h^3}\left(\frac{1}{n_1^2} - \frac{1}{n_2^2}\right) \qquad (17)$$

These correspond to the Balmer Lines of atomic spectra

If n is sufficiently large it could be regarded as a continuous variable and,

$$\frac{d^2 r}{dn^2} = \frac{h^2}{2\pi^2 me^2} \qquad (18)$$

From equations (14, 15) we see,

$$v = \frac{nh}{2\pi mr} = \frac{nh}{2\pi m}\frac{4\pi^2 me^2}{n^2 h^2} = \frac{2\pi e^2}{nh} \qquad (19)$$

Therefore, the de Broglie wavelength associated with the electron in its stationary orbit is given by,

$$\lambda_2 = \frac{h}{mv} = \frac{nh^2}{2\pi me^2} \qquad (20)$$

while its orbital period is given by,

$$T_2 = \frac{2\pi r}{v} = \frac{n^3 h^3}{4\pi^2 me^4} \qquad (21)$$

Eliminating n between equations (20,21) gives,

$$\lambda_2 = \frac{h}{(2\pi m^2 e^2)^{\frac{1}{3}}} T_2^{\frac{1}{3}} \qquad (22)$$

$$\therefore \frac{d\lambda_2}{dT_2} = \frac{h}{3(2\pi m^2 e^2)^{1/3}} T_2^{-2/3} \quad (23)$$

$$\frac{d^2\lambda_2}{dT^2_2} = -\frac{2h}{9(2\pi m^2 e^2)^{1/3}} T_2^{-5/3} \quad (24)$$

But eliminating nh between T and v using equations (19, 21) gives,

$$T_2 = \frac{2\pi e^2}{mv^3} \quad (25)$$

Therefore substituting (25) in equations (23, 24) gives,

$$\frac{d\lambda_2}{dT_2} = \frac{h}{6\pi e^2} v^2 = f_3(c) \quad (26)$$

$$\frac{d^2\lambda_2}{dT^2_2} = -\frac{mh}{18\pi^2 e^4} v^5 = -\frac{mhc^{10}}{18\pi^2 e^4 w^5} = f_4(c) \quad (27)$$

4. Third Derivatives

The Schwarzschild metric for point mass M at rest at the origin O in spherical polar coordinates (r,θ,ϕ) is given by,

$$ds^2 = c^2\gamma dt^2 - \frac{dr^2}{\gamma} - r^2 d\theta^2 - r^2 \sin^2\theta d\phi^2 \quad (28)$$

where,

$$\gamma = 1 - \frac{2GM}{rc^2} \quad (29)$$

For purposes of simplicity consider planar motion in the azimuthal plane $\theta = \frac{\pi}{2}$. In addition, for a light ray, $ds = 0$ and $\phi = cons\tan t = \phi_0$. Equation (28) then reduces to,

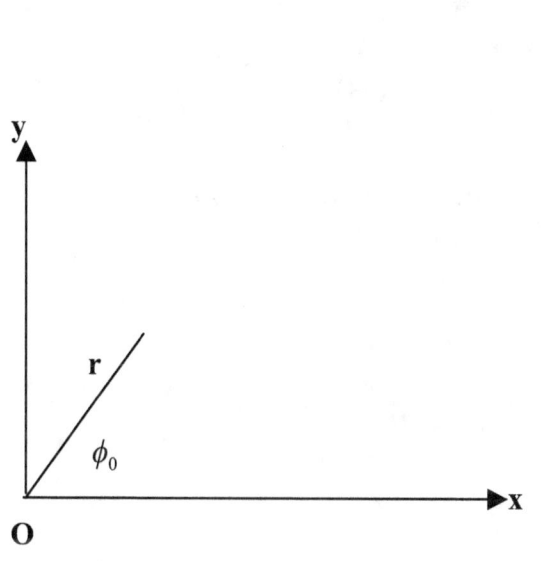
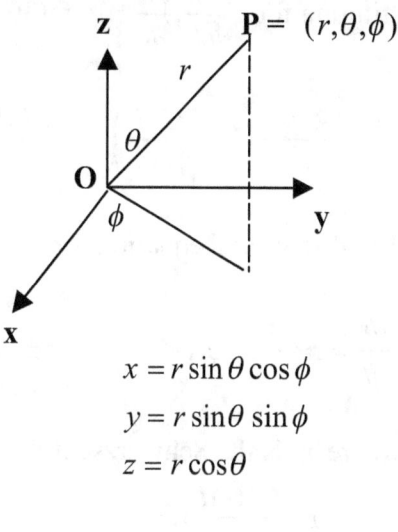

$x = r\sin\theta\cos\phi$
$y = r\sin\theta\sin\phi$
$z = r\cos\theta$

$\boxed{(x,y)\,plane \Leftrightarrow \theta = \pi/2}$

$$\therefore\ (\frac{dr}{dt})^2 = c^2\gamma^2 \quad\Rightarrow\quad \frac{dr}{dt} = \pm c\gamma = \pm c(1 - \frac{2GM}{rc^2}) \qquad (30)$$

Differentiating with respect to time t and using (30) we have,

$$2\frac{dr}{dt}\frac{d^2r}{dt^2} = 2c^2\gamma\frac{d\gamma}{dt}$$

$$\therefore\ \frac{d^2r}{dt^2} = \pm c\frac{d\gamma}{dt} \qquad (31)$$

But from (29),

$$\frac{d\gamma}{dt} = \frac{2GM}{c^2 r^2}\frac{dr}{dt}$$

Using the above result in equations (30, 31),

$$\therefore\ \frac{d^2r}{dt^2} = \pm\frac{2GM}{cr^2}\frac{dr}{dt} = \frac{2GM}{r^2}\left(1 - \frac{2GM}{rc^2}\right) \qquad (32)$$

Differentiating (32),

$$\therefore\ \frac{d^3r}{dt^3} = \pm\frac{2GM}{c}\left(\frac{1}{r^2}\frac{d^2r}{dt^2} - \frac{2}{r^3}(\frac{dr}{dt})^2\right) \qquad (33)$$

Substituting for $\dfrac{d^2r}{dt^2}, \dfrac{dr}{dt}$ from (30, 32) in equation (33) we have,

$$\frac{d^3r}{dt^3} = \mp \frac{4GMc}{r^3}\left(1 - \frac{2GM}{rc^2}\right)\left(1 - \frac{3GM}{rc^2}\right) \tag{34}$$

- The differential equation for $r = r(t)$ is given by the implicit relation (30),

$$\frac{dr}{dt} = \pm c\gamma = \pm c(1 - \frac{2GM}{rc^2}) = \pm c(1 - \frac{r_0}{r})$$

where r_0 is the Schwarzschild radius,

$$r_0 = \frac{2GM}{c^2} \tag{35}$$

$$\therefore \quad dt = \pm \frac{rdr}{c(r-r_0)} = \pm \frac{1}{c}\left(1 + \frac{r_0}{r-r_0}\right)dr$$

Integrating,

$$t = \pm \frac{1}{c}\left[r + r_0 \ln(\frac{|r-r_0|}{r_0})\right] \tag{36}$$

As $r \to 0, t \to \pm \dfrac{1}{c} r_0 \ln r_0, \dfrac{dt}{dr} \to 0$ \hfill (37)

$r \to \infty, t \to \pm\infty, \dfrac{dt}{dr} \to \pm\dfrac{1}{c}$ \hfill (37a)

$r \to r_0-, t \to \mp\infty, \dfrac{dt}{dr} \to \mp\infty$ \hfill (37b)

$r \to r_0+, t \to \mp\infty, \dfrac{dt}{dr} \to \pm\infty$ \hfill (37c)

$\dfrac{dt}{dr} = \pm \dfrac{r}{c(r-r_0)}$ \hfill (37d)

$t = 0 \Rightarrow e^r |r - r_0|^{r_0} = 1$ \hfill (37d)

The $r(t)$ vs. time t curve is shown below

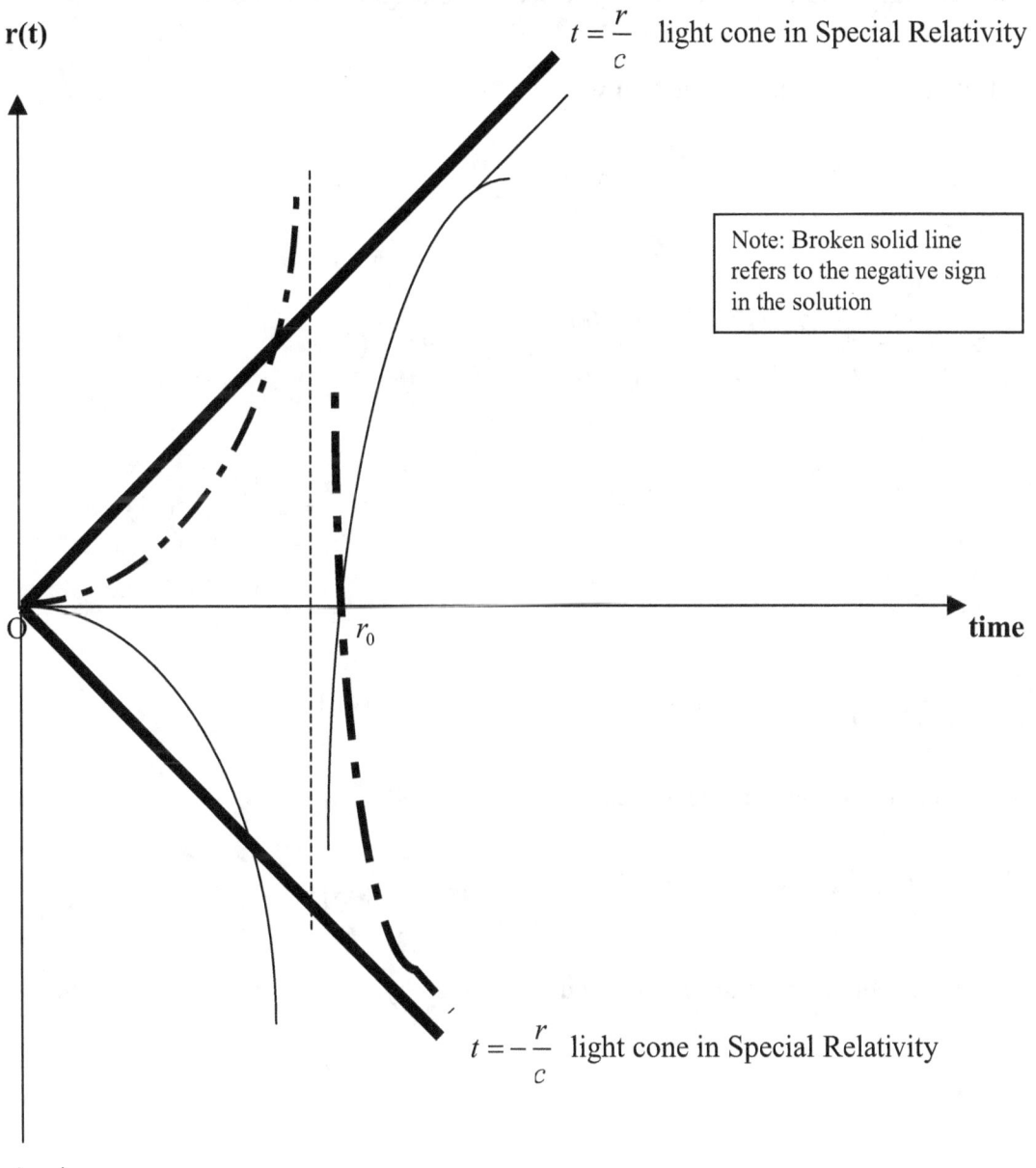

Again,

$$\gamma = 1 - \frac{2GM}{rc^2} = 1 - \frac{r_0}{r} \tag{38}$$

If $r \gg r_0$, $\gamma \approx 1$, and we have the flat space-time approximation. Inside the Schwarzschild radius, $0 < r < r_0$, gravitational effects predominate, while outside the Schwarzschild radius, $r_0 < r < \infty$, the flat space-time approximation becomes valid as $r \to \infty$. At $r = r_0$, there is a
singularity, in the fabric of space-time. Physically, the sphere $r = r_0$ is unreachable.

Inside the Schwarzschild radius, $0 < r < r_0$, close to the origin $r \approx 0$, $\gamma \approx -\dfrac{r_0}{r}$

And from (30, 32, 34) in conjunction with (35),

$$\frac{dr}{dt} \cong \pm c\left(\frac{-2GM}{rc^2}\right) = \mp \frac{2GM}{rc} = \mp \frac{r_0}{r}c \qquad (39)$$

$$\frac{d^2r}{dt^2} \cong \left(\frac{2GM}{r^2}\right)\left(-\frac{2GM}{rc^2}\right) = -\frac{(2GM)^2}{r^3 c^2} = -\frac{r_0^2}{r^3}c^2 \qquad (40)$$

$$\frac{d^3r}{dt^3} \cong \mp\left(\frac{4GMc}{r^3}\right)\left(-\frac{2GM}{rc^2}\right)\left(-\frac{3GM}{rc^2}\right) = \mp \frac{24(GM)^3}{c^3 r^5} = \mp \frac{3r_0^3}{r^5}c^3 \qquad (41)$$

∴ From (40, 41)

$$\frac{d^3r}{dt^3} = \pm \frac{6GM}{r^2 c}\frac{d^2r}{dt^2} \qquad (42)$$

Differentiating (41) with respect to time and using (39)

$$\frac{d^4r}{dt^4} = -\frac{15 r_0^4 c^4}{r^7} \qquad (43)$$

But, the radius of curvature of the $r = r(t)$ curve is given by,

$$\wp = \pm \frac{\left[1 + \left(\dfrac{dr}{cdt}\right)^2\right]^{3/2}}{\dfrac{d^2r}{c^2 dt^2}} \qquad (44)$$

where the factor c is introduced to ensure dimensional compatibility. Using the above approximations (39, 40) for r small (i.e. $\dfrac{1}{r}$ large in comparison to 1),

$$\wp \cong -\frac{2GM}{c^2} \rightarrow \text{circular orbit near the origin, as confirmed in the above sketch.}$$

- The deviation of the metric for curved space-time from the flat space-time metric is measured by

$$\Omega = 1-\gamma = \frac{r_0}{r} \qquad (45)$$

Therefore in the neighborhood of the origin, using the above approximations (39 - 41) for the first, second, and third derivatives we have

$$\frac{d\Omega}{dt} = -r_0\left(\frac{1}{r^2}\frac{dr}{dt}\right) = \pm\frac{r_0^2}{r^3}c \qquad (46)$$

$$\frac{d^2\Omega}{dt^2} = -r_0\left[\frac{1}{r^2}\frac{d^2r}{dt^2} - \frac{2}{r^3}\left(\frac{dr}{dt}\right)^2\right] = \frac{3r_0^3 c^2}{r^5} \qquad (47)$$

$$\frac{d^3\Omega}{dt^3} = -r_0\left[\frac{1}{r^2}\frac{d^3r}{dt^3} - \frac{2}{r^3}\frac{dr}{dt}\frac{d^2r}{dt^2} - \frac{4}{r^3}\frac{dr}{dt}\frac{d^2r}{dt^2} + \frac{6}{r^4}\left(\frac{dr}{dt}\right)^3\right] \qquad (48)$$

$$= -r_0\left[\frac{1}{r^2}\frac{d^3r}{dt^3} - \frac{6}{r^3}\frac{dr}{dt}\frac{d^2r}{dt^2} + \frac{6}{r^4}\left(\frac{dr}{dt}\right)^3\right] = \pm\frac{15r_0^4 c^3}{r^7} \qquad (49)$$

$$\frac{d^4\Omega}{dt^4} = \frac{105 r_0^5 c^4}{r^9} \qquad (50)$$

$$\therefore \quad \frac{d^3\Omega}{dt^3} = \frac{5}{\Omega}\frac{d\Omega}{dt}\frac{d^2\Omega}{dt^2} \qquad (51)$$

$$\frac{d^2\Omega}{dt^2} = \frac{r_0}{r^2}c\frac{d\Omega}{dt} = \frac{\Omega}{r}c\frac{d\Omega}{dt} \qquad (52)$$

Schwarzschild Function, $\gamma(t)$

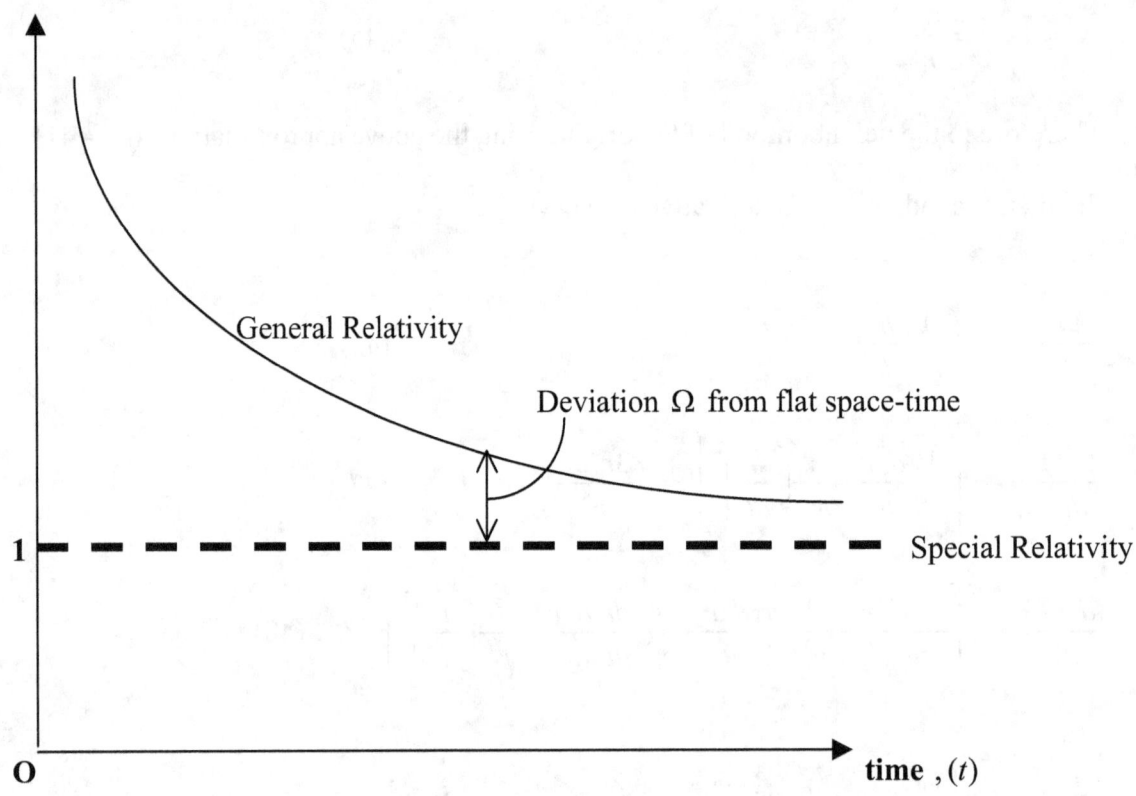

∴ From (51) & (52),

$$\frac{d^3\Omega}{dt^3} = \frac{5c}{r}\left(\frac{d\Omega}{dt}\right)^2 = \frac{5c\Omega}{r_0}\left(\frac{d\Omega}{dt}\right)^2 \qquad (53)$$

Ω is the deviation of curved space-time from flat space-time and is a consequence of the presence of the mass M at the origin. It measures the effect of gravity on the curvature of space-time.

$\dfrac{d\Omega}{dt}$ = a measure of the rate of change of the curvature of space-time, and is indicative of gravitational waves, in much the same manner as the generation of sound waves due to fluctuations in the pressure distribution. The velocity of sound waves v is given by the formula

$$v = \sqrt{\frac{\gamma p}{\rho}} \qquad (54)$$

where,

$$\gamma = \text{ratio of specific heats} = \frac{c_p}{c_v}$$

p, ρ = pressure, and density respectively of the medium in which the sound waves are propagated.

Equation (53) can be rewritten as,

$$\frac{d\Omega}{dt} = \sqrt{\frac{r_0}{5c\Omega} \frac{d^3\Omega}{dt^3}} \quad \Rightarrow \quad r_0 \frac{d\Omega}{dt} = \sqrt{\left(\frac{r_0^3}{5c\Omega}\right) \frac{d^3\Omega}{dt^3}} \qquad (55)$$

Since Ω is a dimensionless variable, $R = r_0 \Omega$ has the dimensions of a length, so that the above equation becomes,

$$V = \frac{dR}{dt} = \sqrt{\frac{rr_0}{5c} \frac{d^3 R}{dt^3}} = \sqrt{\frac{2GMr}{5c^3} \frac{d^3 R}{dt^3}} \qquad (56)$$

Appealing to dimensional analysis,

$$[G] = \frac{MLT^{-2}L^2}{M^2} = M^{-1}L^3 T^{-2} = \frac{T^{-2}}{ML^{-3}} = \frac{T^{-2}}{[\rho]}$$

Defining a variable p which has the dimensions of a pressure - a force per unit area, and a density ρ by the equations,

$$p = \frac{M}{cR} \frac{d^3 R}{dt^3} = \frac{M}{c\Omega} \frac{d^3 \Omega}{dt^3} \qquad (57)$$

$$\rho = \frac{c^2}{Gr_0^2} \rightarrow \text{constant} \qquad (58)$$

Substituting these expressions (57, 58) in equation (56) gives,

$$V = \sqrt{\frac{2p}{5\rho}} \qquad (59)$$

This is analogous to acoustic waves in a medium of constant density ρ. The third derivative of Ω is a measure of the pressure fluctuations in the fabric of curved space-time. It is important to remember these approximations are only valid in the strict interior of the Schwarzschild sphere of radius r_0. Outside the sphere, curved space-time → flat space-time, and hence the quest for gravitational waves becomes problematical.

Since, the gravitational waves travel with the velocity of light c, equation (59) corresponds to an equation of state,

$$p = \frac{5c^2}{2}\rho \tag{60}$$

Therefore, within the Schwarzschild zone the effect of curved space-time can be modeled by a classical gas with an equation of state given by (60). The pressure and density are described by equations (57) and (58). The temperature T is analogous to the particle mass M in r_0 in (58). As the temperature increases (i.e. as r_0 increases) the density decreases. This analogy with the theory of sound is also helpful in addressing certain 2 body problems in general relativity, such as the Doppler effect on the gravitational waves associated with two stars in relative motion. For example, the continuity equation is given by

$$\nabla \bullet (\rho \mathbf{u}) + \frac{\partial \rho}{\partial t} = 0 \tag{61}$$

where $\mathbf{u}$ is the velocity of the fluid, whose density is given by (58), while Euler's equation for non viscous fluids in the presence of external forces is

$$\frac{d}{dt}\mathbf{u} \equiv \frac{\partial}{\partial t}\mathbf{u} + \mathbf{u} \bullet \nabla \mathbf{u} = -\frac{\nabla p}{\rho} + \mathbf{F} \tag{62}$$

where $\mathbf{F}$ is the external force per unit mass with the pressure is given by (57). Integrating the above equation we have along a streamline,

$$\frac{1}{2}\mathbf{u}^2 + \int \frac{dp}{\rho} = \int \mathbf{F} \bullet d\mathbf{r} + \text{constant} \tag{63}$$

If the external force field is conservative, we have

$$\mathbf{F} = -\nabla \phi \tag{64}$$

where ϕ is the potential energy and Bernoulli's equation becomes,

$$\frac{1}{2}\mathbf{u}^2 + \int \frac{dp}{\rho} + \phi = \text{constant} \tag{65}$$

A major advantage of this approach, is the availability of a wide range of the tools of fluid dynamics for handling a class of gravitational problems in general relativity.

5. Fourth Derivatives

For an *inviscid* fluid at rest (hydrostatics), $\mathbf{u} = \mathbf{0}$, or in uniform motion $\frac{d}{dt}\mathbf{u} = 0$, and equation (62) reduces to,

$$\mathbf{F} = -\frac{\nabla p}{\rho} \qquad (66)$$

From (39, 57 & 58), the force **F** acts radially, and

$$\mathbf{F} = -\frac{1}{\rho}\frac{\partial p}{\partial r} = -\frac{1}{\rho}\frac{dp}{dt}\Big/\frac{dr}{dt} = -\frac{1}{\rho}\frac{\frac{M}{c}\frac{d}{dt}\left(\Omega\frac{d^3\Omega}{dt^3}\right)}{\mp\frac{r_0}{r}c}$$

$$= \pm\frac{GMr_0 r}{c^4}\left(\Omega\frac{d^4\Omega}{dt^4} + \frac{d\Omega}{dt}\frac{d^3\Omega}{dt^3}\right) = \pm 120\frac{GMr_0^7}{r^9}$$

$$= \pm 120\frac{GM\Omega^7}{r^2} = \pm\frac{GM}{r^2}(1.98\Omega)^7 \qquad (67)$$

where, $\Omega > 1, \Omega = \frac{r_0}{r}$

A more accurate analysis will require the use of the Navier-Stokes equation for viscous incompressible fluids, given by the equation,

$$\frac{d}{dt}\mathbf{u} \equiv \frac{\partial}{\partial t}\mathbf{u} + \mathbf{u}\bullet\nabla\mathbf{u} = -\frac{\nabla p}{\rho} + \mathbf{F} + \frac{\nu}{\rho}\nabla^2\mathbf{u} \qquad (68)$$

where ν is the coefficient of viscosity.

The Newtonian gravitational field at a distance r from a point mass M is,

$$\Im = \frac{GM}{r^2} \qquad (69)$$

6. Physical Constants

Therefore, within the Schwarzschild sphere, the general relativistic field exceeds the Newtonian value by a factor $\approx 120\Omega^7$. To get a feel for the orders of magnitude involved, consider two particles an electron and a proton at a distance apart of 1 Bohr radius as for example in the case of the hydrogen atom in its ground state. As a second example, at the other extreme consider a white dwarf at its maximum mass, the Chandrasekhar limit of 1.4 solar masses.

- Electron-Proton interaction

$$\text{Bohr radius} = \frac{\hbar^2}{me^2} = 5.29 \times 10^{-9}\,cm. \qquad (70a)$$

Electron charge, $e = 4.80 \times 10^{-10} esu$ (70b)

Proton radius = 1×10^{-13} cm (70c)

Electron rest mass = $9.11 \times 10^{-28} g$ (70d)

Proton rest mass = $1.67 \times 10^{-24} g$ (70e)

Neutron rest mass = $1.67 \times 10^{-24} g$ (70f)

Velocity of light, $c = 3 \times 10^{10} cm/sec$ (70g)

Universal constant of Gravitation, $G = 6.67 \times 10^{-8}$ $cm^3 g^{-1} sec^{-2}$ (70h)

Planck's Constant = 6.63×10^{-27} erg sec (70i)

$\hbar = \dfrac{h}{2\pi} = 1.05 \times 10^{-27}$ erg sec (70i)

Fine structure constant, $\alpha = \dfrac{e^2}{\hbar c} = 1/137$ (70j)

Classical radius of electron = $r_0 = \dfrac{e^2}{m_e c^2} = 2.82 \times 10^{-13}$ cm (70k)

1 atomic unit of time = $\dfrac{r_0}{c} = 9.4 \times 10^{-24}$ sec (70l)

Age of Universe = 18 billion years = 5.68×10^{17} sec (70m)

Mass of Universe = 10^{58} gm (70n)

Size of Universe = 78 billion light years = 7.4×10^{28} cm (70p)

Then, the ratio of

electric force / gravitational force = $\dfrac{e^2}{G m_e m_p} = 2.27 \times 10^{39}$ (70q)

Then the Newtonian force of gravitational attraction is given by;

$F_N = \dfrac{6.67 \times 1.67 \times 9.11 \times 10^{-60}}{27.98 \times 10^{-18}} = 3.63 \times 10^{-42}$ dynes (70r) while the electrostatic force of attraction between the proton and electron is

$$F_E = \frac{23.04 \times 10^{-20}}{27.98 \times 10^{-18}} = 8.23 \times 10^{-3} \quad \text{dynes} \tag{70s}$$

$$\therefore F_E \approx 2.3 \times 10^{39} F_N \tag{70t}$$

while the corresponding general relativistic force within the Schwarzschild sphere ($r \ll r_0$) is given by:

$$r_0 = \frac{2GM}{c^2} = \frac{2 \times 6.67 \times 1.67 \times 10^{-32}}{9 \times 10^{20}} = 2.47 \times 10^{-52} \quad cm \tag{70u}$$

which is inside the nucleus. Take for example,

$$r = \frac{r_0}{2} \Rightarrow \Omega = 2$$

$$F_R = \pm(1.98 \times 2)^7 F_N = \pm 15271 F_N \tag{70v}$$

- Chandrasekhar Limit = 1.4 × solar mass ($M_\Theta = 2 \times 10^{33} g$); This is an upper limit on the mass of a star-- a white dwarf; as the radius decreases the density increases → neutron star → in the limit becomes a black hole.
  Maximum mass of white dwarf = $3 \times 10^{33} g$ (70w)

$$r_0 = \frac{2 \times 6.67 \times 3 \times 10^{25}}{9 \times 10^{20}} = 4.45 \times 10^5 cm = 4.45 \quad km \tag{70x}$$

  A neutron star is essentially a collapsed white dwarf with a mass density $\approx 10^{14} g \; cm^{-3}$ and a radius of about 10 km.

- The fundamental laws of physics are predicated on a few fundamental "constants" of nature such as G the universal constant of gravitation, Planck's constant $\hbar$, and the charge e and rest mass m_0 of the electron. In 1937, Dirac proposed his large numbers hypothesis to explain the huge discrepancy between the electric and gravitational field strengths $\approx 10^{40}$ as we see from our above discussion. He observed this was also the age of the universe in atomic units, using a length scale that corresponds to Bohr's radius. He
  proposed that these so called constants do indeed change with cosmological time and the
  10^{40} factor is a reflection of the age of the universe.

- Some theories speculate that Newton's inverse square law does not apply at the subatomic
  level. Either the value of G needs to be changed or an entirely different law of gravitation
  between subatomic particles need to be employed. We will comment on these proposals
  in Section 9.

7. Quantum Effects

In the presence of a strong gravitational field, such as that encountered at the end of Sections 4 & 5, quantum effects come into play, where elementary particles such as neutrinos are continually being created and destroyed.

Creation Operator : a   number of particles $n \rightarrow n+1$

Destruction Operator : a^+ number of particles $n \rightarrow n-1$

Number Operator : $\mathbf{N} = a\ a^+$; $\mathbf{N}\ |N'\rangle = N'|N'\rangle$

We postulate for our present purposes, these quantized particles form a shell around the mass M which is idealized as a uniform sphere of radius R_0. The $\pm$ sign in front of the gravitational field at the end of Section 4 is due to the possibility of a repulsive effect due to
Quantum effects. For example, the stability of ordinary stars is ensured by the balance between the inward gravitational pull on the star balanced by the outward pressure due to the ionized electron gas surrounding the star and obeying Fermi-Dirac statistics. However, in non-equilibrium states one or other of these competing forces will predominate.
In the Newtonian world space and time are absolute in character and exist independently. This idea was subsequently modified by Einstein in that space and time are intricately linked and one cannot exist without the other. We go a step further, that in the cosmos at large, the electric and magnetic fields cannot exist independently of space-time and conversely. In the
immediate aftermath of the fireball of creation, the electric and magnetic fields separated out to fill the universe (see equation 37).

As $t \rightarrow 0, r \rightarrow 0$, $\Omega \rightarrow 1, \gamma \rightarrow 0$ and $r_0 \rightarrow 0$ such that $\dfrac{r_0}{r} \rightarrow 1$ (71)

This implies that at the moment of the big-bang $t = 0$, $c = \infty$ momentarily, before attaining its normal (constant) value. We postulate that electromagnetism can be linked to Newtonian gravitation without the requirement of the Riemannian calculus of general relativity, where
the curvature tensor $G^{\mu\nu}$ is related to the stress-energy tensor $T^{\mu\nu}$ through the equation,

$$G^{\mu\nu} = -\frac{4\pi G}{c^2} T^{\mu\nu} \qquad (72)$$

where,

$$T^{\mu\nu} = T^{\mu\nu}_{gravity} + T^{\mu\nu}_{hydrodynamics} + T^{\mu\nu}_{electromagnetism} \qquad (73)$$

Some estimates of the large scale features of the electromagnetic fields are described in the next Sections. The deviations from flat space-time in the early universe is partly accounted by the separating out of the electromagnetic fields.

8. Electromagnetism

Again, a Maclaurin series expansion about $t=0$ using values for the derivatives from equations (46 - 50) gives,

$$\Omega(t) = \Omega(0) + t\left(\frac{d\Omega}{dt}\right)_0 + \frac{t^2}{2!}\left(\frac{d^2\Omega}{dt^2}\right)_0 + \frac{t^3}{3!}\left(\frac{d^3\Omega}{dt^3}\right)_0 + \frac{t^4}{4!}\left(\frac{d^4\Omega}{dt^4}\right)_0 + \ldots$$

$$\approx \frac{r_0}{r} \pm \frac{r_0^2}{r^3} c\, t + \frac{t^2}{2}\left(\frac{3r_0^3 c^2}{r^5}\right) \pm \frac{t^3}{6}\left(\frac{15 r_0^4 c^3}{r^7}\right) + \frac{t^4}{24}\left(\frac{105 r_0^5 c^4}{r^9}\right) \pm \ldots \quad (74)$$

where, $r = r(t)$.

Using the Gaussian system of units the dimensions of charge (measured in esu) are given through Coulomb's law in a vacuum,

$$Force = \frac{q_1 q_2}{r^2} \qquad (75)$$

$$\therefore \ [charge] = \left[MLT^{-2}L^2\right]^{\frac{1}{2}} = M^{\frac{1}{2}} L^{\frac{3}{2}} T^{-1}$$

From equation (58) the energy density is given by,

$$W_{gravity} = \frac{c^4}{G r_0^2} \qquad (76)$$

The electrostatic energy in the early universe is given by,

$$W_{electrostatic} = \frac{E^2}{8\pi} \qquad (77)$$

where E is the electric field.

$$[E] = \frac{[charge]}{[distance]} = M^{\frac{1}{2}} L^{\frac{1}{2}} T^{-1}$$

Since part of the gravitational energy is converted into electrostatic energy,

$$W_{gravity} > W_{electrostatic} \qquad (78)$$

$$\therefore \qquad E < \frac{2c^2}{r_0}\sqrt{\frac{2\pi}{G}}$$

This expression gives an upper bound on the electric field. Using numerical values from Section 5,

$$E_{max} \approx \frac{2 \times 9 \times 10^{20}}{4.45 \times 10^5} \sqrt{\frac{2 \times 3.14}{6.67 \times 10^{-8}}} = 4.04 \times 10^{19} \times .97 \approx 4 \times 10^{19} \text{ dynes/esu} \qquad (79)$$

The Lorentz force on a charge q in an electromagnetic filed (**E, H**) is given by,

$$\mathbf{F}_q = q(\mathbf{E} + \frac{1}{c} \mathbf{v} \wedge \mathbf{H}) \qquad (80)$$

Since initially, **v**, $\mathbf{F}_q$ are parallel, i.e. in the radial direction, **H** = 0. Therefore, initially there is no magnetic field.

9. Temperature of early universe:

We compare the equation of state, equation (60) for our hydrodynamic model with the standard equation of state from the kinetic theory of gases,

$$p = \rho k \Theta / m \qquad (81)$$

where,

k = Boltzmann's constant = 1.3805×10^{-16} ergs/degree Centigrade
m = mass of a gas molecule in gm
Θ = temperature of the gas in degrees Kelvin

$$\therefore \quad \frac{5c^2}{2} = \frac{k\Theta}{m} \quad \Rightarrow \Theta = \frac{5c^2 m}{2k} \qquad (82)$$

Assuming the gas is ionized hydrogen,

$$\Theta = \frac{5 \times 9 \times 10^{20} \times (1.67 \times 10^{-24})}{1.38 \times 10^{-16}} \text{ degrees Kelvin}$$

$$= 54 \times 10^{12} \; °K \qquad (83)$$

10. Meson Theory

The modern theory of the atomic nucleus assumes that mesons, particles intermediate in mass
between an electron and a proton are responsible for the nuclear forces. Mesons may carry either a positive, or negative charge, or may even be electrically neutral. According to

Yukawa every proton and neutron in the nucleus continuously emits and reabsorbs mesons. On the other hand, a nearby nucleon might capture an emitted meson from a different nucleon. The associated exchange of momentum in such a transfer constitutes the nuclear force. The constancy in the observed masses of either a proton or neutron is due to the intercession of the uncertainty principle in ensuring conservation of energy is not violated. Nuclear forces are essentially repulsive at short range, while being attractive at large distances. Some of the properties of the meson field are

- It is a charge independent field and its magnitude is the same for proton-proton, neutron-neutron, or proton-neutron interactions
- Short range force
- Strong force and is about 100 times stronger than comparable electromagnetic interactions
- Nuclear force acts only on the immediate neighbors of a given nucleon
- Nuclear force is a non central force i.e. does not act along the straight line joining 2 nucleons
- Nuclear force is spin dependent → nucleons with parallel spins are more strongly bound, than those with anti parallel spin

Although it is now known that mesons are not elementary particles, the above model of the atomic nucleus is still a viable model, especially in its more refined versions. The improved models are more straightforward than the corresponding quark models. They are the basis for many nucleon-nucleon scattering experiments. Quark models are considerably more difficult because they involve 2 to 3 times more particles.

The size of a meson or a nucleon can be estimated using the uncertainty principle connecting time and energy,

$$\Delta t \Delta E \approx \hbar \qquad (84)$$

where,

$$\Delta t \approx r/c, \quad r = \text{range of the nuclear force} \qquad \Delta E \approx m_0 c^2 \qquad (85)$$

$$\therefore \quad r \approx \frac{\hbar}{m_0 c}$$

For a class of mesons with energies in the range 770 - 800 Mev = (770-800) $\times 10^6$ eV = (770-800) $\times 10^6 \times 1.6 \times 10^{-12}$ ergs = (1.3 -1.4) $\times 10^{-24}$ gm, which is less than the mass of the proton. Taking a mean value of 1.35×10^{-24} gm as the rest mass of the meson, we see that the range of the meson is

$$r_m \approx \frac{1.05 \times 10^{-27}}{1.35 \times 10^{-24} \times 3 \times 10^{10}} = 2.6 \times 10^{-14} \text{ cm} \qquad (86)$$

The radius of a meson is less than the radius of a proton (p^+) or neutron (n) or an

electron.
According to quantum field theory,

$$p^+ \rightarrow n + \pi^+ \text{ meson}$$

$$n \rightarrow p^+ + \pi^- \text{ meson}$$

It is important to remember that the above model of the atomic nucleus was proposed in the time period 1935 - 1950, when high energy physics was still in its infancy.
From equation (70x), the Schwarzschild radius for a white dwarf at the Chandrasekhar limit is given by,

$$r_0 = 4.45 \times 10^5 \text{ cm}$$

11. Weak Force

From equations (57, 58, 60) we have

$$\frac{d^3\Omega}{dt^3} = \frac{5c^5}{2MGr_0^2}\Omega \tag{87}$$

This is a 3rd order ordinary differential equation with constant coefficients. Using the operator method ($\frac{d}{dt} \equiv D$), the complementary solution is given by,

$$\Omega = Ae^{x_1 t} + Be^{x_2 t} + Ce^{x_3 t} \tag{88}$$

where A, B, C, are constants of integration and x_1, x_2, x_3 are solutions of the cubic equation

$$D^3 - \frac{5c^5}{2MGr_0^2} = 0 \tag{89}$$

$$x_1 = \left(\frac{5c^5}{2MGr_0^2}\right)^{\frac{1}{3}} \tag{90}$$

$$x_2 = \frac{1}{2}\left(\frac{5c^5}{2MGr_0^2}\right)^{\frac{1}{3}}(-1 + i\sqrt{3}) \tag{90a}$$

$$x_3 = \frac{1}{2}\left(\frac{5c^5}{2MGr_0^2}\right)^{\frac{1}{3}}\left(-1-i\sqrt{3}\right) \tag{90b}$$

These last 2 solutions correspond to damped harmonic oscillations with period given by,

$$T_3 = \frac{4\pi}{\sqrt{3}}\left(\frac{2MGr_0^2}{5c^5}\right)^{\frac{1}{3}}$$

$$= \frac{8\pi GM}{3^{\frac{1}{2}}5^{\frac{1}{3}}c^3} \tag{90c}$$

From Section 4, we see that Bernoulli's equation (65) for constant ρ gives in conjunction with equation (60)

$$\frac{1}{2}\mathbf{u}^2 + \frac{p}{\rho} - \frac{GM}{r} = \text{constant} = \frac{1}{2}\mathbf{u}^2 + \frac{5c^2}{2} - \frac{GM}{r}$$

This implies that
$$\frac{1}{2}\mathbf{u}^2 - \frac{GM}{r} = \text{constant.} \tag{91}$$

The corresponding de Broglie wavelength is,

$$\lambda_3 = \frac{h}{Mu^*} \tag{92}$$

where u^* is the particle velocity defined by

$$u^* = \frac{c^2}{u} = \mp c\frac{r}{r_0}$$

using

$$u = \frac{dr}{dt} = \mp \frac{r_0}{r}c \text{, from equation (39)}$$

Since $r < r_0$, $u > c$. This corresponds to a shock wave akin to the phenomenon of Cerenkov
radiation in the fabric of space time and is the velocity of the associated wave packet. Taking
numerical values we have,

$$\lambda_3 = \frac{hr_0}{Mcr} > \frac{h}{Mc} \tag{93}$$

To probe small distances of an atomic scale we require high energy projectiles with as small a value of λ_3 as possible. Therefore we take λ_3 to be

$$\lambda_3 = \frac{h}{Mc}\kappa \tag{94}$$

where $\kappa = \frac{r_0}{r} > 1$ is a non dimensional constant scaling factor. Optimally, κ should be as small as possible consistent with $\kappa > 1$. Eliminating M between equations (90c) & (94)

$$\lambda_3 = \alpha T_3^{-1} \tag{95}$$

where,

$$\alpha = \frac{8\pi Gh}{3^{\frac{1}{2}} 5^{\frac{1}{3}} c^4} \kappa = 4.63 \times 10^{-75} \kappa \quad cm-\sec \tag{96}$$

$$\therefore \frac{d\lambda_3}{dT_3} = -\alpha T_3^{-2} \tag{97a}$$

$$\frac{d^2\lambda_3}{dT_3^2} = 2\alpha T_3^{-3} \tag{97b}$$

$$\frac{d^3\lambda_3}{dT_3^3} = -6\alpha T_3^{-4} \tag{97c}$$

We shall now calibrate this relationship to obtain the weak force, in a manner applicable to both the subatomic particles within an atomic nucleus, and to the galaxies in the cosmos. Using the method of dimensions,

$$\left[\frac{d^3\lambda_3}{dT_3^3}\right] = \left[\frac{Force}{Mass}\right]\left[\frac{Velocity}{Length}\right] = \left[LT^{-3}\right] \tag{98a}$$

$$[\alpha] = [LT] \tag{98b}$$

$$\therefore [Force] = \left[\frac{Length}{Velocity}\right][Mass]\left[\frac{d^3\lambda_3}{dT_3^3}\right] \tag{98c}$$

The Coulomb electrostatic force between 2 protons in the nucleus has the dimension

$$[F_{el}] = \frac{[e^2]}{[L^2]} \tag{98d}$$

where e is the electronic charge. Consider a non dimensional variable defined by,

$$\left[\frac{Force}{F_{el}}\right] = \frac{\Re^3}{ce^2} \aleph \left[\frac{d^3 \lambda_3}{dT_3^3}\right] = -6\alpha T_3^{-4} \frac{\Re^3}{ce^2} \aleph = -4 \times 10^{-66} T_3^{-4} \Re^3 \aleph \kappa \tag{99}$$

where $\Re, \aleph$ are length and mass factors respectively to be appropriately selected, where

$length \sim \Re$; $Mass \sim \aleph$; $Velocity = c$

Since the weak force is about 10^{-7} the strength of the electromagnetic force, we postulate,

$$4 \times 10^{-66} T_3^{-4} \Re^3 \aleph \kappa = 10^{-7} \tag{100}$$

∴ The period of the matter waves is
∴

$$T_3 = (40 \Re^3 \aleph \kappa)^{\frac{1}{4}} \times 10^{-15} \quad s \tag{101}$$

and the particle velocity is

$$u^* = \frac{c}{\kappa} \leq c \quad cm/s \tag{102}$$

The de Broglie formula gives,

$$\lambda_3 = \frac{h}{Mc} \kappa \rightarrow 4.63 \times 10^{-75} \kappa T_3^{-1} = \frac{h\kappa}{Mc} \tag{103}$$

∴ $$M = 4.8 \times 10^{22} \times (40 \Re^3 \aleph \kappa)^{\frac{1}{4}} \tag{104a}$$

Therefore the ratio of the particle mass M to the rest mass of an electron is given by,

$$\frac{M}{m_e} = 5.3 \times 10^{49} \times (40 \Re^3 \aleph \kappa)^{\frac{1}{4}} \tag{104b}$$

If the rest mass of the particle is M_0 then,

$$M = \frac{M_0}{\sqrt{1 - \frac{u^{*2}}{c^2}}} = \frac{M_0 \kappa}{\sqrt{\kappa^2 - 1}}$$

$$\therefore \quad \frac{M_0}{m_e} = 5.3 \times 10^{49} \times \left(40 \Re^3 \aleph \kappa\right)^{\frac{1}{4}} \times \frac{\sqrt{\kappa^2 - 1}}{\kappa} \tag{105}$$

Let,

$\Re$ = Schwarzschild radius for an electron = $\frac{2 G m_e}{c^2} = 1.34 \times 10^{-55}$ cm (106a)

$\aleph$ = rest mass of electron = m_e (106b)

We believe an electron is an appropriate entity to introduce as it belongs to the class of leptons which are not affected by the strong force. In addition our methodology brings gravitational effects into the mix through the constant G. Since the function,

$$f(\kappa) = \frac{\kappa^{\frac{1}{4}} \sqrt{\kappa^2 - 1}}{\kappa} = \frac{\sqrt{\kappa^2 - 1}}{\kappa^{\frac{3}{4}}} \quad ; \quad \kappa > 1 \tag{106c}$$

is monotonic increasing, the smallest possible value of κ is,

$$\kappa = 1 + \varepsilon \quad ; \quad \varepsilon \to 0+ \tag{106d}$$

Take $\kappa = 1.05$, corresponding to particles moving at .95 times the speed of light. Then from (105),

$$\frac{M_0}{m_e} \approx 48 \tag{107}$$

This probably is a neutral tau neutrino, which has a maximum mass ratio of about 53. Higher values of κ will result in lower mass ratios. It should be kept in mind that electrons and tau neutrinos belong to the lepton family of particles, which are affected only by the gravitational and the weak force.

$f(\kappa)$

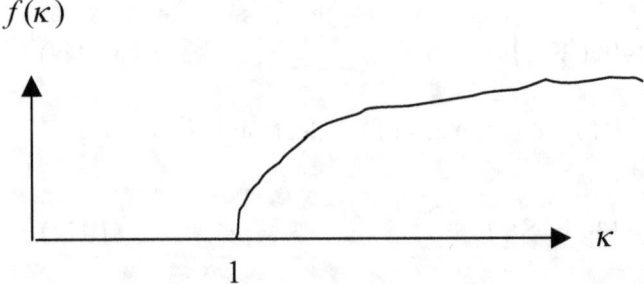

12. Strong Force

From (97c),

$$\frac{d^4\lambda_3}{dT_3^4} = 24\alpha T_3^{-5} \tag{108}$$

and following the strategy of Section 11, the strong force is hypothesized to be

$$[Force_{str}] = \left[\frac{Length}{Velocity}\right]^2 [Mass]\left[\frac{d^4\lambda_3}{dT_3^4}\right] = [MLT^{-2}] \tag{109a}$$

$$\therefore \left[\frac{Force_{str}}{F_{el}}\right] = \frac{\mathfrak{R}_1^4 \aleph_1}{c^2 e^2}\left[\frac{d^4\lambda_3}{dT_3^4}\right] = \frac{24\alpha\mathfrak{R}_1^4 \aleph_1}{c^2 e^2} T_3^{-5} \tag{109b}$$

Since the strong force is 20 times the electromagnetic force,

$$\frac{24\alpha\mathfrak{R}_1^4 \aleph_1}{c^2 e^2} T_3^{-5} = 20 \tag{110}$$

where $\mathfrak{R}_1, \aleph_1$ are length and mass factors to be appropriately selected. Therefore, the period of the oscillations is,

$$T_3 \to T_4 = (.0266 \mathfrak{R}_1^4 \aleph_1 \kappa)^{\frac{1}{5}} \times 10^{-15} \text{ sec} \tag{111}$$

$$\lambda_3 \to \lambda_4 \tag{111a}$$

From (103), the associated particle mass is

$$M = 4.8 \times 10^{22} \times (.0266 \mathfrak{R}_1^4 \aleph_1 \kappa)^{\frac{1}{5}} \tag{112}$$

Analogous to equation (105), the rest mass ratio is,

$$\frac{M_0}{m_e} = 5.3 \times 10^{49} \times (.0266 \mathfrak{R}_1^4 \aleph_1 \kappa)^{\frac{1}{5}} \times \frac{\sqrt{\kappa^2 - 1}}{\kappa} \tag{113}$$

Let us take $\kappa = 1.05$ as before and

$$\mathfrak{R}_1 = \text{Schwarzschild radius for a proton} = 2.47 \times 10^{-52} \ cm \quad (114a)$$

$$\aleph_1 = \text{rest mass of a proton} = m_p \quad (115)$$

Substituting,

$$\frac{M_0}{m_e} \approx 348 \quad (116)$$

which is probably a pion $\pi^{\pm}$ but with a somewhat larger mass than expected.

13. Standard Model

The primary purpose of this endeavor is to explore the essential unity between the 4 fundamental forces of nature as we presently understand them. The strong force which holds the protons and neutrons in a reasonably stable mode is mediated by gluons. Some of the highlights of the Standard Model are shown in the flowchart below and include :

- Fundamental particles = leptons [electron, neutrinos, tau (τ) and the muon (μ) particles], quarks, photons, and $W^{\pm}, Z^0$ particles. They are indivisible, while the elementary particles are composites of these
- The quarks in the proton and neutron are held together by gluons. Stable nucleii correspond to relatively light quark masses, while nucleii with heavy quarks are unstable.
- Masses of the $W^{\pm}, Z^0$ particles are greater than the proton mass, by a factor of between 81 to 92
- Up quark (u) has a charge = $\frac{2}{3}e$, while the down quark (d) has a charge = $-\frac{1}{3}e$. The mass of the down quark is about 2 times the mass of the up quark
- Mesons are bosonic hadrons, while protons and neutron are fermionic hadrons.
- In addition to the up & down quarks, there are 4 other types or flavours of quarks of differing mass → top(t), bottom (b), strange (s), and charm (c). For example omega-minus, Ω^- is a baryon made of 3 strange quarks = (sss), while the lambda particle is the baryon Λ = (uds), both being somewhat more massive than the proton.
- Equation (83) estimates the temperature of the universe in the immediate aftermath of the big-bang based on our model. This would correspond to a window of about $[10^{-6}, 10^{-4}]$ seconds after the initial creation saga according to present astrophysical consensus. During this time interval the initial soup of gluons, quarks, electrons and neutrinos coalesced to form protons and neutrons.

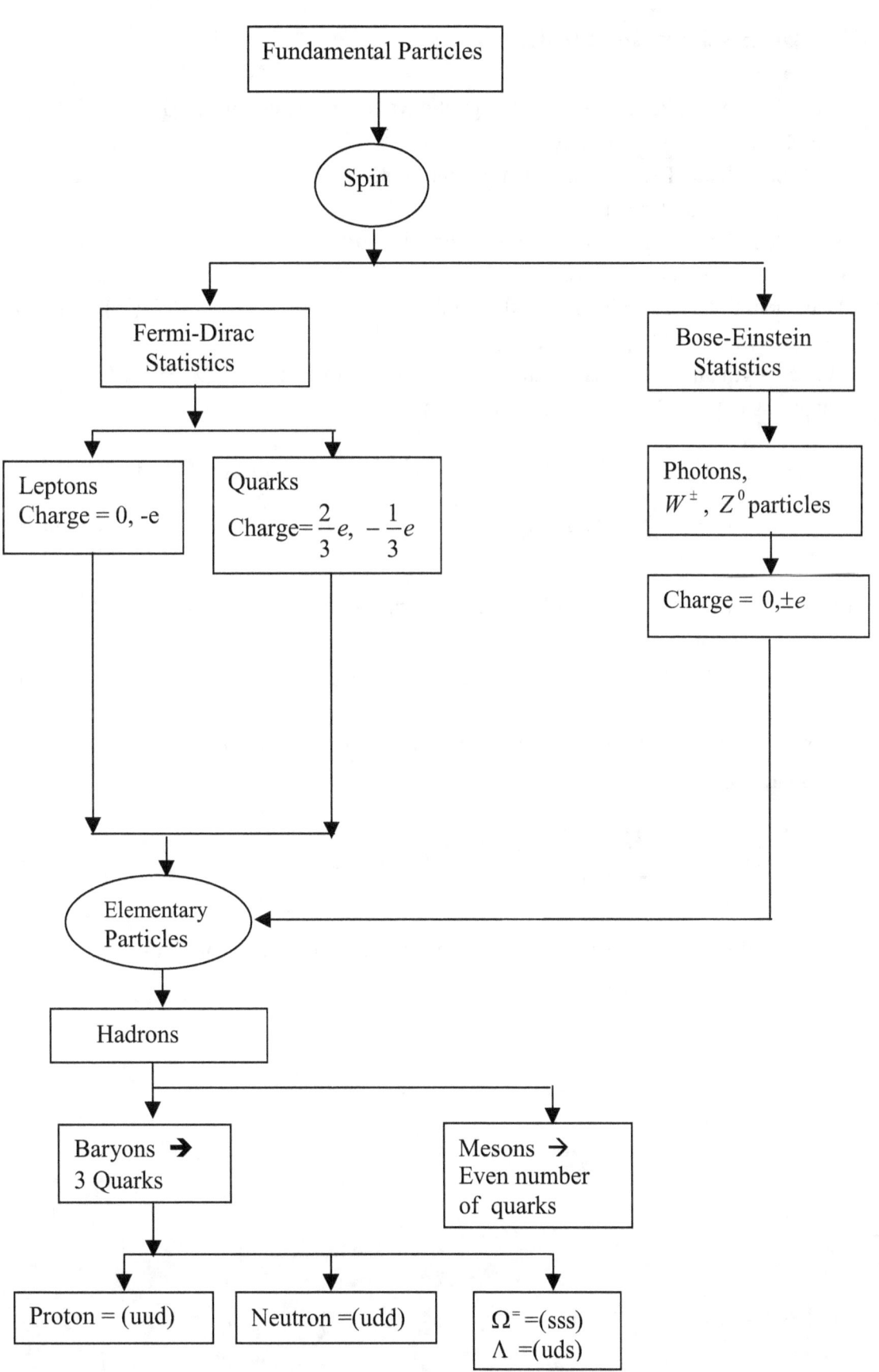

14. Superposition Principle

- Wave length and Period of electromagnetic waves in a vacuum : $\lambda_0, T_0; \lambda_0 = \lambda_0(T_0)$, equations (2) → first derivatives
- Wave length and Period of classical orbital motion : $\lambda_1, T_1; \lambda_1 = \lambda_1(T_1)$, equations (6,7,8) → second derivatives
- Wave length and Period of quantum mechanical orbits : $\lambda_2, T_2; \lambda_2 = \lambda_2(T_2)$, equations (22, 23, 24) → second derivatives
- Wave length and Period associated with the weak force : $\lambda_3, T_3; \lambda_3 = \lambda_3(T_3)$, equations (95, 96, 97a, 97b, 97c;) → third derivatives
- Wave length and Period associated with the strong force : $\lambda_4, T_4; \lambda_4 = \lambda_4(T_4)$, equations (111, 111a) → fourth derivatives

$$\lambda(T_1, T_2, T_3, T_4, \ldots) = c_0 \lambda_0(T_0) + c_1 \lambda_1(T_1) + c_2 \lambda_2(T_2) + c_3 \lambda_3(T_3) + c_4 \lambda_4(T_4) + \ldots$$
$$= \sum_{k=0}^{\infty} c_k \lambda_k(T_k) \qquad (117)$$

where, $c_0, c_1, c_2, c_3, c_4, \ldots$ are probabilistic weighting factors such that

$$|c_0|^2 + |c_1|^2 + |c_2|^2 + |c_3|^2 + |c_4|^2 + \ldots = 1 \qquad (118)$$

and each of the functions $\lambda_k(T_k); k \geq 0$ is represented by their Maclaurin series approximations,

$$\lambda_k(T_k) = \sum_{r=0}^{\infty} \frac{T_k^r}{r!} \left[\frac{d^r \lambda_k}{dT_k^r} \right]_0 \quad ; \quad k = 0,1,2,3,4,\ldots \qquad (119)$$

the subscript 0 in (119) corresponding to time =0, the instant of the big-bang.

Chapter 5 Ultra Relativistic Motion of Charged Particles

> " Science enhances the moral values of life because it furthers a love of truth and reverence—love of truth displaying itself in the constant endeavor to arrive at a more exact knowledge of the world of mind and matter around us , and reverence, because every advance in knowledge brings us face to face with the mystery of our own being."
>
> Max Planck, Nobel Laureate

Abstract :

This chapter studies the ultra relativistic motion of a charged particle in an electromagnetic field. The shock waves generated are analogous in concept to transonic flow regimes in aerodynamics. The electromagnetic mass of the particle ,assumed to be spherical in shape, changes as the light barrier is approached. Consequently, charge invariance may be violated. We assume for purposes of simplicity, that the motion is rectilinear.

Radiation Damping

The radiation field due to a point charge q is given by equations (24, 25) of the solutions to question (10) in the WPD chapter,

$$\vec{E}(\vec{r},t) = \frac{-q[\vec{a}' - (\vec{a}' \bullet \vec{\varepsilon}')\vec{\varepsilon}'] + q\vec{\varepsilon}' \wedge \left(\vec{a}' \wedge \frac{\vec{v}'}{c}\right)}{c^2 |\vec{r} - \vec{r}'| \left(1 - \frac{\vec{v}' \bullet \vec{\varepsilon}'}{c}\right)^3} = q\frac{\vec{\varepsilon}' \wedge \left((\vec{\varepsilon}' - \frac{\vec{v}'}{c}) \wedge \vec{a}'\right)}{c^2 |\vec{r} - \vec{r}'| \left(1 - \frac{\vec{v}' \bullet \vec{\varepsilon}'}{c}\right)^3} \quad (1)$$

$$\vec{H}(\vec{r},t) = \vec{\varepsilon}' \wedge \frac{-q\vec{a}' + q\vec{\varepsilon}' \wedge \left(\vec{a}' \wedge \frac{\vec{v}'}{c}\right)}{c^2 |\vec{r} - \vec{r}'| \left(1 - \frac{\vec{v}' \bullet \vec{\varepsilon}'}{c}\right)^3} = \vec{\varepsilon}' \wedge \vec{E} \quad (2)$$

and,

$$E^2(\vec{r},t) = H^2(\vec{r},t) \quad (3)$$

The Poynting vector is given by,

$$\vec{S}(\vec{r},t) = \frac{c}{4\pi}\vec{E}(\vec{r},t) \wedge \vec{H}(\vec{r},t) = \frac{c}{4\pi}E^2\vec{\varepsilon}'$$

$$= \frac{q^2}{4\pi c^3 |\vec{r}-\vec{r}'|^2} \left\{ \frac{2(\vec{\varepsilon}'\bullet\vec{a}')(\vec{v}'\bullet\vec{a}')}{c\left(1-\frac{\vec{v}'\bullet\vec{\varepsilon}'}{c}\right)^5} + \frac{\vec{a}'^2}{\left(1-\frac{\vec{v}'\bullet\vec{\varepsilon}'}{c}\right)^4} - \frac{\left(1-\frac{\vec{v}'^2}{c^2}\right)(\vec{\varepsilon}'\bullet\vec{a}')^2}{\left(1-\frac{\vec{v}'\bullet\vec{\varepsilon}'}{c}\right)^6} \right\} \vec{\varepsilon}' \qquad (4)$$

Therefore, the rate at which energy is radiated by the charge is obtained by integrating the expression (4) over a sphere of radius $= |\vec{r}-\vec{r}'|$ centered at the point $\vec{r}'$.

$$\frac{dE}{dt} = \iint (\vec{S}\bullet\vec{\varepsilon}') |\vec{r}-\vec{r}'|^2 \sin\theta d\theta d\phi$$

$$= \frac{q^2}{2c^3} \int \left\{ \frac{2(\vec{\varepsilon}'\bullet\vec{a}')(v'\bullet\vec{a}')}{c\left(1-\frac{\vec{v}'\bullet\vec{\varepsilon}'}{c}\right)^5} + \frac{\vec{a}'^2}{\left(1-\frac{\vec{v}'\bullet\vec{\varepsilon}'}{c}\right)^4} - \frac{\left(1-\frac{\vec{v}'^2}{c^2}\right)(\vec{\varepsilon}'\bullet\vec{a}')^2}{\left(1-\frac{\vec{v}'\bullet\vec{\varepsilon}'}{c}\right)^6} \right\} d\theta \qquad (5)$$

To compute the total energy radiated throughout the motion of the particle we must integrate (5) over the time, remembering that the integrand is a function of t'.

$$t' = t - \frac{|\vec{r}-\vec{r}'|}{c} \qquad (6)$$

Differentiating with respect to t',

$$1 = \frac{dt}{dt'} + \frac{(\vec{r}-\vec{r}')\bullet\vec{v}'}{c|\vec{r}-\vec{r}'|} \quad \Rightarrow \quad \frac{dt}{dt'} = 1 - \frac{\vec{\varepsilon}'\bullet\vec{v}'}{c} \qquad (7)$$

Therefore from (5, 7) we have,

$$E = \frac{q^2}{2c^3} \int \int \left\{ \frac{2(\vec{\varepsilon}'\bullet\vec{a}')(\vec{v}'\bullet\vec{a}')}{c\left(1-\frac{\vec{v}'\bullet\vec{\varepsilon}'}{c}\right)^4} + \frac{\vec{a}'^2}{\left(1-\frac{\vec{v}'\bullet\vec{\varepsilon}'}{c}\right)^3} - \frac{\left(1-\frac{\vec{v}'^2}{c^2}\right)(\vec{\varepsilon}'\bullet\vec{a}')^2}{\left(1-\frac{\vec{v}'\bullet\vec{\varepsilon}'}{c}\right)^5} \right\} d\theta dt' \qquad (8)$$

We now perform an integration by parts in the time interval t_1', t_2', assuming that the state of motion of the particle is the same at the 2 time extremities. Remembering, $\vec{a}' = \frac{d\vec{v}'}{dt'}$,

$$\vec{v}' = \frac{d\vec{r}'}{dt'}$$

$$\int \left\{ \frac{2(\vec{\varepsilon}' \bullet \vec{a}')(\vec{v}' \bullet \vec{a}')}{c\left(1 - \frac{\vec{v}' \bullet \vec{\varepsilon}'}{c}\right)^4} + \frac{\vec{a}'^2}{\left(1 - \frac{\vec{v}' \bullet \vec{\varepsilon}'}{c}\right)^3} - \frac{\left(1 - \frac{\vec{v}'^2}{c^2}\right)(\vec{\varepsilon}' \bullet \vec{a}')^2}{\left(1 - \frac{\vec{v}' \bullet \vec{\varepsilon}'}{c}\right)^5} \right\} dt'$$

$$= -\int_{t_1}^{t_2} \left\{ \frac{d}{dt'} \left[\frac{2(\vec{\varepsilon}' \bullet \vec{a}')\vec{v}'}{c\left(1 - \frac{\vec{v}' \bullet \vec{\varepsilon}'}{c}\right)^4} + \frac{\vec{a}'}{\left(1 - \frac{\vec{v}' \bullet \vec{\varepsilon}'}{c}\right)^3} - \frac{\left(1 - \frac{\vec{v}'^2}{c^2}\right)(\vec{\varepsilon}' \bullet \vec{a}')\vec{\varepsilon}'}{\left(1 - \frac{\vec{v}' \bullet \vec{\varepsilon}'}{c}\right)^5} \right] \bullet \vec{v}' \right\} dt'$$

$$= = -\int_{t_1}^{t_2} \left\{ \frac{d}{dt'} \left[\frac{2(\vec{\varepsilon}' \bullet \vec{a}')\vec{v}'}{c\left(1 - \frac{\vec{v}' \bullet \vec{\varepsilon}'}{c}\right)^4} + \frac{\vec{a}'}{\left(1 - \frac{\vec{v}' \bullet \vec{\varepsilon}'}{c}\right)^3} - \frac{\left(1 - \frac{\vec{v}'^2}{c^2}\right)(\vec{\varepsilon}' \bullet \vec{a}')\vec{\varepsilon}'}{\left(1 - \frac{\vec{v}' \bullet \vec{\varepsilon}'}{c}\right)^5} \right] \right\} \bullet d\vec{r}' \qquad (9)$$

Therefore, the reaction force of the radiation is given by,

$$\vec{F}_r = \frac{q^2}{2c^3} \int_0^\pi d\theta \frac{d}{dt'} \left[\frac{2(\vec{\varepsilon}' \bullet \vec{a}')\vec{v}'}{c\left(1 - \frac{\vec{v}' \bullet \vec{\varepsilon}'}{c}\right)^4} + \frac{\vec{a}'}{\left(1 - \frac{\vec{v}' \bullet \vec{\varepsilon}'}{c}\right)^3} - \frac{\left(1 - \frac{\vec{v}'^2}{c^2}\right)(\vec{\varepsilon}' \bullet \vec{a}')\vec{\varepsilon}'}{\left(1 - \frac{\vec{v}' \bullet \vec{\varepsilon}'}{c}\right)^5} \right]$$

$$= \frac{q^2}{2c^3} \frac{d}{dt'} \int_0^\pi \left[\frac{2(\vec{\varepsilon}' \bullet \vec{a}')\vec{v}'}{c\left(1 - \frac{\vec{v}' \bullet \vec{\varepsilon}'}{c}\right)^4} + \frac{\vec{a}'}{\left(1 - \frac{\vec{v}' \bullet \vec{\varepsilon}'}{c}\right)^3} - \frac{\left(1 - \frac{\vec{v}'^2}{c^2}\right)(\vec{\varepsilon}' \bullet \vec{a}')\vec{\varepsilon}'}{\left(1 - \frac{\vec{v}' \bullet \vec{\varepsilon}'}{c}\right)^5} \right] d\theta$$

$$= \frac{q^2}{2c^3} \frac{d}{dt'} \int_0^\pi \left[\frac{2a'\cos\theta \vec{v}'}{c\left(1 - \frac{v'\cos\theta}{c}\right)^4} + \frac{\vec{a}'}{\left(1 - \frac{v'\cos\theta}{c}\right)^3} - \frac{\left(1 - \frac{\vec{v}'^2}{c^2}\right)a'\cos\theta \vec{\varepsilon}'}{\left(1 - \frac{v'\cos\theta}{c}\right)^5} \right] d\theta \qquad (10)$$

where θ is the angle between $\vec{\varepsilon}', \vec{v}'$ or $\vec{a}'$. Equation (10) is derived on the assumption that the charged particle is moving on a straight line.

General Solutions.

In this Section we show in some detail the analytical procedures, for obtaining closed form solutions for the integrals in (10).
Consider for instance the integral,

$$I_n = \int_0^\pi \frac{\cos\theta}{(1-\alpha\cos\theta)^n} d\theta \quad \text{,where } n = \text{integer}, \quad 0 \leq \alpha = \frac{v'}{c} \leq 1 \qquad (11)$$

Put, $t = \tan\frac{\theta}{2}$, $\beta^2 = \frac{1-\alpha}{1+\alpha}$ \hfill (11a)

$$\therefore \quad I_n = \int_0^\infty \frac{\frac{1-t^2}{1+t^2}}{\left(1-\alpha\frac{1-t^2}{1+t^2}\right)^n} \frac{2dt}{1+t^2} = \frac{2}{(1+\alpha)^n} \int_0^\infty \frac{(1-t^2)(1+t^2)^{n-2}}{(\beta^2+t^2)^n} dt \qquad (12)$$

But,

$$I = \int_0^\infty \frac{(1-t^2)(1+t^2)^{n-2}}{(\beta^2+t^2)^n} dt = \int_0^\infty \frac{(1-t^2)\sum_{r=0}^{n-2} {}^{n-2}C_r t^{2r}}{(t^2+\beta^2)^n} dt = \int_0^\infty \frac{\sum_{r=0}^{n-2} {}^{n-2}C_r t^{2r} - \sum_{r=0}^{n-2} {}^{n-2}C_r t^{2r+2}}{(t^2+\beta^2)^n} dt$$

$$= \int_0^\infty \frac{\sum_{r=0}^{n-2} {}^{n-2}C_r t^{2r} - \sum_{r=1}^{n-1} {}^{n-2}C_{r-1} t^{2r}}{(t^2+\beta^2)^n} dt = \int_0^\infty \frac{1-t^{2n-2} + \sum_{r=1}^{n-2} a_r t^{2r}}{(t^2+\beta^2)^n} dt \qquad (13)$$

where,

$$a_r = {}^{n-2}C_r - {}^{n-2}C_{r-1}$$

$$= \frac{(n-2)!(n-2r-1)}{r!(n-r-1)!} \qquad (13a)$$

$$\therefore \quad I = \int_0^\infty \frac{1-t^{2n-2}}{(t^2+\beta^2)^n} dt + \sum_{r=1}^{n-2} a_r \int_0^\infty \frac{t^{2r}}{(t^2+\beta^2)^n} dt \qquad (13b)$$

Let,

$$I(n,r) = \int_0^\infty \frac{t^{2r}}{(t^2+\beta^2)^n} dt = \frac{1}{2}\int_0^\infty t^{2r-1} \frac{2t}{(t^2+\beta^2)^n} dt, \quad 1 \leq r \leq n-2 \qquad (13c)$$

Integrating by parts, we obtain the recurrence relation

$$I(n,r) = \frac{2r-1}{2(n-1)} I(n-1, r-1)$$

$$\dots\dots\dots\dots\dots\dots\dots\dots\dots\dots\dots$$

$$= \frac{1}{2^r} \frac{(2r-1)(2r-3)\dots 1}{(n-1)(n-2)\dots(n-r)} I(n-r,0)$$

$$= \frac{(2r)!(n-r-1)!}{2^{2r} r!(n-1)!} \cdot \frac{1}{\beta^{2n-2r-1}} \int_0^{\frac{\pi}{2}} \cos^{2(n-r-1)}\theta \, d\theta \qquad (13c1)$$

To complete (13c) we use the standard result for integral N

$$\int_0^{\frac{\pi}{2}} \cos^{2N}\theta \, d\theta = \frac{2N!}{2(2^N N!)^2} \pi \qquad (13c2)$$

To evaluate the first integral in (13b) we use partial fractions in conjunction with the following Lemma.

Lemma : Consider the quotient of 2 polynomials $P(x), Q(x)$ where the degree of $P(x)$ is less than the degree of $Q(x)$. If the equation $Q(x) = 0$ has a multiple root $x = r$ of multiplicity m, then we can write the quotient in terms of partial fractions

$$\frac{P(x)}{Q(x)} = \frac{P(x)}{(x-r)^m q(x)} = \frac{A_1}{(x-r)} + \frac{A_2}{(x-r)^2} + \dots + \frac{A_m}{(x-r)^m} + \dots$$

successive applications of Leibnitz's Rule,

$$(uv)_n = uv_n + {}^n C_1 u_1 v_{n-1} + {}^n C_2 u_2 v_{n-2} + \dots + u_n v$$

give the recurrence relations,

$$P(r) = A_m q(r)$$

$$P'(r) = A_m q'(r) + A_{m-1} q(r) \qquad (13d)$$

$$P^{(m-1)}(r) = A_m q^{(m-1)}(r) + (m-1)A_{m-1} q^{(m-2)}(r) + (m-1)(m-2)A_{m-2} q^{(m-3)}(r) + \dots$$

$$+ \dots + (m-1)! A_1 q(r)$$

Complex roots occur in conjugate pairs, with corresponding conjugate numerators.

We now use the results (13d) from above Lemma to evaluate the following integrals needed to compute the expression (10).

Case (1):

$$n = m = 4, \quad r = i\beta,$$

$$\text{Let,} \quad I = \int_0^\infty \frac{1-t^6}{(t^2+\beta^2)^4} dt = \int_0^\infty \frac{1-t^6}{(t-i\beta)^4(t+i\beta)^4} dt \tag{14}$$

Decomposing the integrand in (14) into partial fractions,,

$$\frac{1-t^6}{(t-i\beta)^4(t+i\beta)^4} \equiv \frac{A_1}{(t-i\beta)} + \frac{A_2}{(t-i\beta)^2} + \frac{A_3}{(t-i\beta)^3} + \frac{A_4}{(t-i\beta)^4} + \ldots$$

where in accordance with decomposition Lemma,

$$P(t) = 1 - t^6, \quad q(t) = (t+i\beta)^4 \; ; \; P(i\beta) = 1 + \beta^6, \; q(i\beta) = 16\beta^4$$

$$P'(t) = -6t^5, \quad P'(i\beta) = -6i\beta^5$$

$$P''(t) = -30t^4, \quad P''(i\beta) = -30\beta^4$$

$$P^{(3)} = -120t^3, \quad P^{(3)}(i\beta) = 120i\beta^3$$

$$q'(t) = 4(t+i\beta)^3, \quad q'(i\beta) = 4(i\beta+i\beta)^3 = -32i\beta^3$$

$$q''(t) = 12(t+i\beta)^2 \; ; \; q''(i\beta) = 12(i\beta+i\beta)^2 = -48\beta^2$$

$$q'''(t) = 24(t+i\beta) \; ; \; q'''(i\beta) = 24(i\beta+i\beta) = 48i\beta$$

Therefore from (13d),

$$A_4 = \frac{1+\beta^6}{16\beta^4}$$

$$A_3 = i\frac{1-2\beta^6}{8\beta^5} \tag{15}$$

$$A_2 = -\frac{5+11\beta^6}{32\beta^6}$$

$$A_1 = i\frac{15(-1+\beta^6)}{96\beta^7}$$

Therefore,

$$\frac{1-t^6}{(t-i\beta)^4(t+i\beta)^4} \equiv \sum_{k=1}^{4}\frac{A_k}{(t-i\beta)^k}+\frac{A_k^*}{(t+i\beta)^k} \tag{15a}$$

where (*) denotes complex conjugation.

Case (2)) :

$n = m = 5$, $r = i\beta$,

Let, $$I = \int_0^{\infty}\frac{1-t^8}{(t^2+\beta^2)^5}dt = \int_0^{\infty}\frac{1-t^8}{(t-i\beta)^5(t+i\beta)^5}dt \tag{16}$$

Similar to the procedures adopted for Case (1),

$$\frac{1-t^8}{(t-i\beta)^5(t+i\beta)^5} \equiv \frac{A_1}{(t-i\beta)}+\frac{A_2}{(t-i\beta)^2}+\frac{A_3}{(t-i\beta)^3}+\frac{A_4}{(t-i\beta)^4}+\frac{A_5}{(t-i\beta)^5}+\ldots$$

with,

$P(t) = 1-t^8$, $q(t) = (t+i\beta)^5$; $P(i\beta) = 1-\beta^8$, $q(i\beta) = 32i\beta^5$

$P'(t) = -8t^7$, $P'(i\beta) = 8i\beta^7$

$P''(t) = -56t^6$, $P''(i\beta) = 56\beta^6$

$P^{(3)} = -336t^5$, $P^{(3)}(i\beta) = -336i\beta^5$

$P^{(4)}(t) = -1680t^4$; $P^{(4)}(i\beta) = -1680\beta^4$

$q'(t) = 5(t+i\beta)^4$, $q'(i\beta) = 5(i\beta+i\beta)^4 = 80\beta^4$

$q''(t) = 20(t+i\beta)^3$; $q''(i\beta) = 20(i\beta+i\beta)^3 = -160i\beta^3$

$q'''(t) = 60(t+i\beta)^2$; $q'''(i\beta) = 60(i\beta+i\beta)^2 = -240\beta^2$

$q^{(4)}(t) = 120(t+i\beta)$; $q^{(4)}(i\beta) = 120(i\beta+i\beta) = 240i\beta$

Therefore,

$$A_5 = -i\frac{1-\beta^8}{32\beta^5}$$

$$A_4 = \frac{5+11\beta^8}{64\beta^6}$$

$$A_3 = i\frac{15-47\beta^8}{128\beta^7} \qquad (16a)$$

$$A_2 = \frac{-35-93\beta^8}{256\beta^8}$$

$$A_1 = -i\frac{35(1-\beta^8)}{256\beta^9}$$

Therefore,

$$\frac{1-t^6}{(t-i\beta)^4(t+i\beta)^4} \equiv \sum_{k=1}^{4} \frac{A_k}{(t-i\beta)^k} + \frac{A_k^*}{(t+i\beta)^k}$$
(16b)

Case (3) :

$$J_n = \int_0^\pi \frac{1}{(1-\alpha\cos\theta)^n} d\theta = \frac{2}{(1+\alpha)^n} \int_0^\infty \frac{(1+t^2)^{n-1}}{(\beta^2+t^2)^n} dt \qquad (17)$$

Take, $n = m = 3$, with

$$\frac{(1+t^2)^2}{(t^2+\beta^2)^3} \equiv \sum_{k=1}^{3} \frac{A_k}{(t-i\beta)^k} + \frac{A_k^*}{(t+i\beta)^k} \; ;$$
(17a)

$$P(t) = (1+t^2)^2 = 1+2t^2+t^4 \; ; \; r = i\beta \; ; \; q(t) = (t+i\beta)^3$$

$$P(i\beta) = 1-2\beta^2+\beta^4 \; ; \; q(i\beta) = -8i\beta^3$$

$$P'(t) = 4t+4t^3 \qquad ; \; P'(i\beta) = 4i\beta+4i^3\beta^3 = i4(\beta-\beta^3)$$

$$P''(t) = 4+12t^2 \qquad ; \; P''(i\beta) = 4-12\beta^2$$

$$P'''(t) = 24t \qquad ; \; P'''(i\beta) = 24i\beta$$

$$q'(t) = 3(t+i\beta)^2 \quad ; \quad q'(i\beta) = 3(i\beta + i\beta)^2 = -12\beta^2$$

$$q''(t) = 6(t+i\beta) \quad ; \quad q''(i\beta) = 6(i\beta + i\beta) = 12i\beta$$

Therefore,

$$A_3 = i\frac{1 - 2\beta^2 + \beta^4}{8\beta^3}$$

$$A_2 = \frac{-3 - 2\beta^2 + 5\beta^4}{16\beta^4} \tag{17b}$$

$$A_1 = -i\frac{3 + 4\beta^2 + 3\beta^4}{16\beta^5}$$

Therefore to evaluate the integrals: → (14) in conjunction with (15, 15a) ; (16) in conjunction with (16a, 16b) ; & (17) in conjunction (17a, 17b) we use elementary standard results. For example, (15, 16a, 17b) show that A_1 is pure imaginary, therefore

$$A_1 = ia_1 \,, \text{ where } a_1 \text{ is real.}$$

Therefore for $k=1$, we have from (15a, 16b, 17a),

$$\frac{A_1}{t - i\beta} + \frac{A_1^*}{t + i\beta} = -\frac{2a_1\beta}{t^2 + \beta^2} \Rightarrow \int_0^\infty \left(\frac{A_1}{t - i\beta} + \frac{A_1^*}{t + i\beta}\right) dt = -a_1\pi \tag{17c}$$

Ultra Relativistic Solutions

When the particle is travelling at speeds near the speed of light, we have from (11, 11a),

$$\alpha \approx 1 \,, \quad \beta^2 \approx 0 \Rightarrow \beta^2 = \text{small}$$

and we can approximate (15, 16a, 17b) by respectively

Case(1):

$$A_4 = \frac{1}{16\beta^4}$$

$$A_3 = i\frac{1}{8\beta^5} \tag{18a}$$

$$A_2 = -\frac{5}{32\beta^6}$$

$$A_1 = -i\frac{15}{96\beta^7} = -i\frac{5}{32\beta^7}$$

Case (2):

$$A_5 = -i\frac{1}{32\beta^5}$$

$$A_4 = \frac{5}{64\beta^6}$$

$$A_3 = i\frac{15}{128\beta^7} \quad (18b)$$

$$A_2 = \frac{-35}{256\beta^8}$$

$$A_1 = -i\frac{35}{256\beta^9}$$

Case (3):

$$A_3 = i\frac{1-2\beta^2}{8\beta^3}$$

$$A_2 = \frac{-3-2\beta^2}{16\beta^4}$$

(18c)

$$A_1 = -i\frac{3+4\beta^2}{16\beta^5}$$

4. Ultra Relativistic Particle Motion

Therefore, the equation of motion is,

$$\frac{d}{dt}\left(\frac{m_0\vec{v}}{\sqrt{1-\frac{v^2}{c^2}}}\right) = q\left(\vec{E}(\vec{r},t) + \frac{1}{c}\vec{v}\wedge H(\vec{r},t)\right) - \vec{F}_r \qquad (19)$$

where from (10, 11, 17) the radiation damping term becomes,

$$\vec{F}_r = \frac{q^2}{2c^3}\frac{d}{dt}\left[\frac{2aI_3}{c}\vec{v} + J_3\vec{a} + (1-\frac{v^2}{c^2})aI_5\vec{\varepsilon}\right] \qquad (19a)$$

where for convenience we have omitted the prime in the integration variable. From (12, 13a, 13b, 14, 18a; 16, 18b),

$$I_3 = \frac{2}{(1+\alpha)^3}\int_0^\infty \frac{1-t^6}{(t^2+\beta^2)^4}dt$$

$$= \frac{2}{(1+\alpha)^3}\int_0^\infty \frac{-5i}{32\beta^7(t-i\beta)} - \frac{5}{32\beta^6(t-i\beta)^2} + \frac{i}{8\beta^5(t-i\beta)^3} + \frac{1}{16\beta^4(t-i\beta)^4}dt +$$

$$\frac{2}{(1+\alpha)^3}\int_0^\infty \frac{5i}{32\beta^7(t+i\beta)} - \frac{5}{32\beta^6(t+i\beta)^2} - \frac{i}{8\beta^5(t+i\beta)^3} + \frac{1}{16\beta^4(t+i\beta)^4}dt \qquad (19b)$$

Integrating and simplifying,

$$I_3 = \frac{5\pi}{16\beta^7(1+\alpha)^3} \qquad (19c)$$

$$I_5 = \frac{2}{(1+\alpha)^5}\left[\int_0^\infty \frac{1-t^8}{(t^2+\beta^2)^5}dt + \sum_{r=1}^3 a_r \int_0^\infty \frac{t^{2r}}{(t^2+\beta^2)^5}dt\right] \qquad (19d)$$

But,

$$\frac{2}{(1+\alpha)^5}\int_0^\infty \frac{1-t^8}{(t^2+\beta^2)^5}dt$$

$$= \frac{2}{(1+\alpha)^5}\int_0^\infty dt\left[\begin{array}{c}\frac{-35i}{256\beta^9(t-i\beta)} - \frac{35}{256\beta^8(t-i\beta)^2} + \frac{15i}{128\beta^7(t-i\beta)^3} + \\ \frac{5}{64\beta^6(t-i\beta)^4} - \frac{i}{32\beta^5(t-i\beta)^5}\end{array}\right] +$$

$$\frac{2}{(1+\alpha)^5} \int_0^\infty dt \left[\frac{35i}{256\beta^9(t+i\beta)} - \frac{35}{256\beta^8(t+i\beta)^2} - \frac{15i}{128\beta^7(t+i\beta)^3} + \frac{5}{64\beta^6(t+i\beta)^4} + \frac{i}{32\beta^5(t+i\beta)^5} \right]$$

$$= \frac{35\pi}{128\beta^9(1+\alpha)^5} \tag{19e}$$

and from (13a, 13b, 13c),

$$\sum_{r=1}^3 a_r \int_0^\infty \frac{t^{2r}}{(t^2+\beta^2)^5} dt = a_1 I(5,1) + a_2 I(5,2) + a_3 I(5,3) = 2[I(5,1) - I(5,3)]$$

$$= \frac{5\pi}{128}\left(\frac{1}{\beta^7} - \frac{1}{\beta^3}\right) \tag{19f}$$

Therefore, substituting (19e, 19f) in (19d),

$$I_5 = \frac{5\pi}{128(1+\alpha)^5}\left(\frac{7}{\beta^9} + \frac{1}{\beta^7} - \frac{1}{\beta^3}\right)$$

(19g)

From (17, 17a, 18c)

$$J_3 = \frac{2}{(1+\alpha)^3} \int_0^\infty dt \left[\frac{-i(3+4\beta^2)}{16\beta^5(t-i\beta)} + \frac{-3-2\beta^2}{16\beta^4(t-i\beta)^2} + \frac{i(1-2\beta^2)}{8\beta^3(t-i\beta)^3} + \frac{i(3+4\beta^2)}{16\beta^5(t+i\beta)} + \frac{-3-2\beta^2}{16\beta^4(t+i\beta)^2} - \frac{i(1-2\beta^2)}{8\beta^3(t+i\beta)^3} \right]$$

$$= \frac{(3+4\beta^2)\pi}{8\beta^5(1+\alpha)^3} \tag{19h}$$

Therefore using (19c, 19g, 19h) equation (19a) for the radiation damping which is a frictional force becomes

$$\vec{F}_r = \frac{\pi q^2}{16c^3}\frac{d}{dt}\left\{\frac{1}{\beta^3(1+\alpha)^3}\left[\frac{5a}{8\beta^4 c}\vec{v} + \frac{3+4\beta^2}{\beta^2}\vec{a} + \frac{5(1-\frac{v^2}{c^2})a}{16(1+\alpha)^2}\left(\frac{7}{\beta^6}+\frac{1}{\beta^4}-1\right)\vec{\varepsilon}\right]\right\}_{t=t'}$$

$$= \frac{\pi q^2}{128 c^3}\frac{d}{dt}\left\{\frac{(1+\beta^2)^3}{\beta^6}\left[\frac{5a}{8\beta^4 c}\vec{v} + \frac{3+4\beta^2}{\beta^2}\vec{a} + \frac{5\beta^2 a}{16}\left(\frac{7}{\beta^6}+\frac{1}{\beta^4}-1\right)\vec{\varepsilon}\right]\right\}_{t=t'} \qquad (19i)$$

As the velocity of the particle approaches the velocity of light, i.e. $\beta \to 0$, we see that the damping due to the electromagnetic field approaches infinity.

4. Electromagnetic Mass

The energy of the electromagnetic field is given by,

$$W = \frac{1}{8\pi}\int_R^\infty\int_0^\pi\int_0^{2\pi}(E^2+H^2)s^2\sin\theta\,ds\,d\theta\,d\phi \qquad (20)$$

where, $s = |\vec{r}-\vec{r}'|$ is the magnitude of the retarded radius vector and we assume the particle is modeled by a sphere of radius R. In addition, the Coulomb type part of the electric vector is,

$$\vec{E} = q\frac{\left(\vec{\varepsilon}'-\frac{\vec{v}'}{c}\right)\left(1-\frac{v'^2}{c^2}\right)}{s^2\left(1-\frac{\vec{v}'\bullet\vec{\varepsilon}'}{c}\right)^3} \qquad (21)$$

$$\vec{H} = \vec{\varepsilon}' \wedge \vec{E} \qquad (22)$$

Therefore,

$$W = \frac{q^2}{8\pi R}\int_0^\pi\int_0^{2\pi}\left\{\frac{\left(1-\frac{v'^2}{c^2}\right)^2\left[\left(\vec{\varepsilon}'-\frac{\vec{v}'}{c}\right)^2+\left(\frac{\vec{v}'\wedge\vec{\varepsilon}'}{c}\right)^2\right]}{\left(1-\frac{\vec{v}'\bullet\vec{\varepsilon}'}{c}\right)^6}\right\}\sin\theta\,d\theta\,d\phi$$

$$= \frac{q^2}{4R}\left(1-\frac{v'^2}{c^2}\right)^2 \int_0^\pi \frac{1+2\frac{v'^2}{c^2}-\frac{2v'\cos\theta}{c}-\frac{v'^2\cos^2\theta}{c^2}}{\left(1-\frac{v'\cos\theta}{c}\right)^6}\sin\theta\, d\theta$$

Put $t = \cos\theta$

$$\therefore\ W = \frac{q^2 c^6}{4R v'^6}\left(1-\frac{v'^2}{c^2}\right)^2 \int_{-1}^1 \frac{1+2\frac{v'^2}{c^2}-\frac{2v'␣t}{c}-\frac{v'^2 t^2}{c^2}}{\left(\frac{c}{v'}-t\right)^6}\, dt \tag{23}$$

It should be noted that the expression (20) for the electromagnetic field energy is not Lorentz invariant, because the surface of the moving charge is not spherical to an observer at rest. As before, decompose the integrand into partial fractions,

$$P(t) = (1+2\alpha^2) - 2\alpha t - \alpha^2 t^2\ ;\ r = \gamma = \frac{c}{v'} = \frac{1}{\alpha}\ ;\ q(t) = 1 \tag{24}$$

$$\frac{P(t)}{(\gamma - t)^6} = \sum_{k=1}^6 \frac{A_k}{(t-\gamma)^6} \tag{24a}$$

$\therefore$ Using procedures similar to those that led to (13d),

$$(1+2\alpha^2) - 2\alpha t - \alpha^2 t^2 \equiv A_6 + A_5(t-\gamma) + A_4(t-\gamma)^2 + A_3(t-\gamma)^3 + A_2(t-\gamma)^4 + A_1(t-\gamma)^5$$

$\therefore\ P(\gamma) = A_6 \Rightarrow A_6 = -2(1-\alpha^2)$

$$P'(\gamma) = A_5 \Rightarrow A_5 = -4\alpha \tag{24b}$$
$$P''(\gamma) = 2A_4 \Rightarrow A_4 = -\alpha^2$$

$$P'''(\gamma) = P^{(4)}(\gamma) = P^{(5)}(\gamma) = 0 \Rightarrow A_3 = A_2 = A_1 = 0$$

Using (24, 24a, 24b) in (23) gives,

$$W = \frac{q^2}{4R\alpha^6}(1-\alpha^2)^2 \int_{-1}^1 \frac{-2(1-\alpha^2)}{(t-\gamma)^6} - \frac{4\alpha}{(t-\gamma)^5} - \frac{\alpha^2}{(t-\gamma)^4}\, dt$$

$$= \frac{q^2(-\alpha^4 + 10\alpha^2 + 15)}{30R(1-\alpha^2)^2} \tag{25}$$

But the particle energy is given by,

$$W = \frac{m_0 c^2}{\sqrt{1-\alpha^2}} \tag{25a}$$

Therefore, from (25, 25a), the radius of the charged particle is given by,

$$R = \frac{q^2(-\alpha^4 + 10\alpha^2 + 15)}{30 m_0 c^2 (1-\alpha^2)^{3/2}} \tag{26}$$

Therefore as $v \to c-0$, i.e as $\alpha \to 1$, $R \to \infty$, unless $q \to 0$. This implies that as the velocity of the particle approaches the speed of light, it morphs either into an electrically neutral particle such as a neutrino, or a photon. Conservation of charge requires that the charge on the particle is either transmitted to its surroundings or a particle of different rest mass is created with a charge = q.

Q and A Section for Ch 1

1. Show that $\int_{-\infty}^{+\infty} e^{-x^2} dx = \sqrt{\pi}$

2. Calculate the classical partition function for (a) the kinetic energy of N non interacting particles each of mass m, in a container of volume V (b) N harmonic oscillators in 3 dimensions with the same frequency.. Hence, calculate the mean energy of the systems at temperature T. Comment on how these results relate to the equipartition of energy.

3(a). Use an appropriate partition function to calculate the specific heat of a solid in which the atoms are treated as independent oscillators with the same frequency. The atoms are further assumed to vibrate about fixed lattice points. (b) Derive Planck's Law for black body radiation. (c) Derive Stefan-Boltzmann Law → the energy density of black body radiation is proportional to the 4th power of the absolute temperature.

4. Calculate the energy levels for the hydrogen atom using the relativistic Sommerfeld-Bohr elliptical orbits.

5. Consider a classical dynamical system of n degrees of freedom, defined by generalized coordinates $q_1, q_2, ..., q_n$ and generalized momenta $p_1, p_2, ..., p_n$. Let F, G be any arbitrary functions of the q's and the p's.

$$F = F(q_1, q_2, ..., q_n; p_1, p_2, ..., p_n, t)$$
$$G = G(q_1, q_2, ..., q_n; p_1, p_2, ..., p_n, t)$$

Then the Poisson Bracket of F & G is defined by,

$$\{F, G\} = \sum_{k=1}^{k=n} \left(\frac{\partial F}{\partial q_k} \frac{\partial G}{\partial p_k} - \frac{\partial F}{\partial p_k} \frac{\partial G}{\partial q_k} \right)$$

Show that:
$$\{q_r, q_s\} = 0$$
$$\{p_r, p_s\} = 0$$
$$\{q_r, p_s\} = \delta_{rs}$$

6. (a) Derive Dirac's equation for a free particle (b) Show that it corresponds to particles with spin = $\pm \frac{\hbar}{2}$.
   (c) Write Dirac's equation for an electron in an electromagnetic field.

7. Under a suitable Gauge Condition show that the electromagnetic potentials $\vec{A}(r,t), \phi(\vec{r},t)$ satisfy the equations

$$\nabla^2 \mathbf{A} - \frac{1}{c^2} \frac{\partial^2}{\partial t^2} \mathbf{A} = -\frac{4\pi}{c} \mathbf{j}$$

$$\nabla^2 \phi - \frac{1}{c^2}\frac{\partial^2}{\partial t^2}\phi = -4\pi\rho$$

Hence show that,

$$\phi(\vec{r},t) = \int_{allspace} \frac{\rho\left(\vec{r}',t \pm \frac{|\vec{r}-\vec{r}'|}{c}\right)}{|\vec{r}-\vec{r}'|} dV'$$

$$\vec{A}(\vec{r},t) = \frac{1}{c}\int_{allspace} \frac{\vec{j}\left(\vec{r}',t \pm \frac{|\vec{r}-\vec{r}'|}{c}\right)}{|\vec{r}-\vec{r}'|} dV'$$

8. Use Maxwell's Equations to derive the Biot-Savart Law for the magnetic field due to a current carrying line element

   Hence or otherwise show that the force between 2 wires 1 & 2 of arbitrary shape and length L_1, L_2 and carrying currents i_1, i_2 respectively is given by the vector line integral

$$\frac{1}{c^2}i_1 i_2 \oiint_{L_1, L_2} \frac{d\vec{s}_2 \wedge [d\vec{s}_1 \wedge (\vec{r}_2 - \vec{r}_1)]}{|\vec{r}_2 - \vec{r}_1|^3}$$

9. Using the results of exercise 7 derive the Lienard-Wiechert potentials for a charge q, moving with a velocity $\vec{v}'(T)$

$$\phi(\vec{r},t) = \frac{q}{|\vec{r}-\vec{r}'|\left(1 - \frac{\vec{v}'(T)\bullet\vec{\varepsilon}'(T)}{c}\right)}$$

$$\vec{A}(\vec{r},t) = \frac{q\vec{v}'(T)}{c|\vec{r}-\vec{r}'|\left(1 - \frac{\vec{v}'(T)\bullet\vec{\varepsilon}'(T)}{c}\right)}$$

where,

$$T = t \pm \frac{|\vec{r}-\vec{r}'(T)|}{c}$$

$$\vec{\varepsilon}' = \frac{\vec{r}-\vec{r}'(T)}{|\vec{r}-\vec{r}'|}$$

10. Use the results of exercise 9 to determine the radiation field. Hence, show that $\vec{E}(\vec{r},t), \vec{H}(\vec{r},t)$ for the radiation field are orthogonal to each other and are each perpendicular to $\vec{\varepsilon}'$.

Solutions :

1. Let, $I = \int_{-\infty}^{+\infty} e^{-x^2} dx$

    $\therefore I^2 = \left(\int_{-\infty}^{+\infty} e^{-x^2} dx \right) \left(\int_{-\infty}^{+\infty} e^{-y^2} dy \right) = \iint_{-\infty}^{\infty} e^{-(x^2+y^2)} dxdy$

    Put,

    $x = r\cos\theta$
    $y = r\sin\theta$

    Therefore, the Jacobian of the transformation is given by

    $$\frac{\partial(x,y)}{\partial(r,\theta)} = r$$

    $\therefore dxdy = \frac{\partial(x,y)}{\partial(r,\theta)} drd\theta = rdrd\theta$

    $\therefore I^2 = \int_0^{+\infty} e^{-r^2} rdr \int_0^{2\pi} d\theta = \pi \Rightarrow I = \sqrt{\pi}$

2. The mean energy of the system at a temperature T is given by

    $$<E> = -\frac{\partial}{\partial \beta} \ln Z$$

    where,

    $$\beta = \frac{1}{kT}; \quad k = \text{Boltzmann's Constant}$$

    (a). The Hamiltonian for a free particle of mass m is,

    $$H = \frac{p^2}{2m} = \frac{p_x^2 + p_y^2 + p_z^2}{2m}$$

    where p is the particle momentum. Therefore, the partition function for a *single particle* is,

    $$\Xi = \iiint dxdydz \iiint e^{-\frac{p_x^2+p_y^2+p_z^2}{2mkT}} dp_x dp_y dp_z = \frac{V}{h^3} \left(\int_{-\infty}^{+\infty} e^{-\frac{p_x^2}{2mkT}} dp_x \right) \left(\int_{-\infty}^{+\infty} e^{-\frac{p_y^2}{2mkT}} dp_y \right) \left(\int_{-\infty}^{+\infty} e^{-\frac{p_z^2}{2mkT}} dp_z \right)$$

    $$= V(2\pi mkT)^{\frac{3}{2}} / h^3 \quad (V = \text{volume of container})$$

using the result from exercise 1

$$\int_{-\infty}^{+\infty} e^{-\lambda t^2} dt = \sqrt{\frac{\pi}{\lambda}}$$

For N particles, the classical partition function is

$$Z = \frac{1}{N!} \Xi^N = \frac{V^N}{N!} (2\pi mkT)^{\frac{3N}{2}} / h^{3N}$$

$$<E> = \frac{3N}{2\beta} = \frac{3}{2} NkT$$

This is the translational energy, and does not include the energy associated with the internal degrees of freedom. The energy per translational degree of freedom $= \frac{1}{2} kT$ as stipulated in the equipartition of Energy.

2(b) For a harmonic oscillator in 3 dimensions with the usual notation,

$$\mathbf{F} = -\omega^2 \mathbf{r} \quad ; \quad \mathbf{F} = -\nabla V$$

These equations imply a potential energy V

$$-\omega^2 r = -\frac{dV}{dr} \quad \Rightarrow V = \frac{1}{2} \omega^2 r^2$$

Therefore, the Hamiltonian is given by

$$H = \frac{p^2}{2m} + \frac{1}{2} \omega^2 r^2 = \frac{p_x^2 + p_y^2 + p_z^2}{2m} + \frac{1}{2} \omega^2 (x^2 + y^2 + z^2)$$

∴ $$\Xi = \frac{1}{h^3} \iiint \iiint e^{-\frac{H}{kT}} dxdydzdp_x dp_y dp_z = \frac{1}{h^3} \iiint e^{-\frac{p^2}{2mkT}} dp_x dp_y dp_z \iiint e^{-\frac{\omega^2 r^2}{2kT}} dxdydz$$

∴ using the results of exercise 2(a) we have,

$$\Xi = \frac{1}{h^3} (2\pi mkT)^{\frac{3}{2}} \left(\frac{2\pi kT}{\omega^2}\right)^{\frac{3}{2}} = m^{\frac{3}{2}} \left(\frac{2\pi kT}{\omega h}\right)^3$$

Therefore for N oscillators,

$$Z = \frac{1}{N!} \Xi^N = \frac{m^{\frac{3}{2}N}}{N!} \left(\frac{2\pi kT}{\omega h}\right)^{3N}$$

$$<E> = 3NkT$$

using the first law of thermodynamics

$$\delta Q = dE(T) + pdV$$

We have for the specific heat at constant volume

$$c_v = \left.\frac{\delta Q}{\partial T}\right|_V = \left.\frac{\partial E}{\partial T}\right|_V = 3Nk \quad \rightarrow \text{Dulong \& Petit's Law}$$

3(a). According to quantum mechanics, the energy of a harmonic oscillator is given by,

$$E_r = h\nu\left(r + \frac{1}{2}\right) \quad ; \quad r = 0,1,2,3,.... \quad (1)$$

Therefore, the partition function for a single oscillator is

$$\Xi = \sum_{r=0}^{\infty} e^{-\beta E_r} \quad ; \quad \beta = \frac{1}{kT} \quad (2)$$

substituting (1) in (2) and simplifying,

$$\Xi = e^{-\frac{x}{2}} \sum_{r=0}^{\infty} e^{-xr} = \frac{e^{-\frac{x}{2}}}{1 - e^{-x}} \quad (3)$$

where,

$$x = \beta h\nu = \frac{h\nu}{kT} \quad (4)$$

Therefore, from (2)

$$\Xi = \frac{e^{-\frac{x}{2}}}{1 - e^{-x}} \quad (5)$$

$$\therefore \quad \ln \Xi = -\frac{x}{2} - \ln(1 - e^{-x})$$

$$\therefore \quad <E> = -\frac{\partial}{\partial \beta}\ln \Xi = \frac{h\nu}{2} + \frac{h\nu}{e^{\frac{h\nu}{kT}} - 1} \quad (6)$$

Einstein Model: From result 2(b), the total energy is given by,

$$E = 3N<E> \quad (7)$$

specific heat at constant volume is

$$C_V = \left(\frac{\delta Q}{\partial T}\right)_V = \left(\frac{\partial E}{\partial T}\right)_V = 3Nk\frac{x^2 e^x}{(e^x-1)^2} \qquad (8)$$

3(b) Consider a photon gas of N non-interacting identical particles. In the state specified by the occupation numbers

$$n_1, n_2, ..., n_r, ... \qquad (9)$$

it has an energy,

$$E(n_1, n_2, n_3, ..., n_r, ..) = \sum_r n_r \varepsilon_r \qquad (10)$$

where the summation is over all single particle states $r = 1,2,3,..,$. Since photons satisfy the Bose-Einstein Statistics,

$$n_r = 0,1,2,3,... \quad \text{for all } r \qquad (11)$$

Therefore, the partition function is given by

$$Z(T,V) = \sum_{n_1,n_2,} e^{-\beta E} = \sum_{n_1=0}^{\infty}\sum_{n_2=0}^{\infty}.....e^{-\beta(n_1\varepsilon_1+n_2\varepsilon_2+....)} \qquad (12)$$

It should be borne in mind that the occupation numbers $n_1, n_2, ..., n_r, ...$ assume all possible values $0,1,2,3,...$ independently of the values of the other occupation numbers, because there is no restriction on the total number of particles.
Therefore (12) reduces to,

$$Z(T,V) = \prod_{r=1}^{r=\infty} \frac{1}{1-e^{-\beta\varepsilon_r}} \qquad (13)$$

Therefore,

$$\ln Z(T,V) = -\sum_{r=1}^{r=\infty} \ln(1-e^{-\beta\varepsilon_r}) \qquad (14)$$

and the mean occupation number corresponding to the average number of photons in the state - r,

$$<n_r> = -\frac{1}{\beta}\left(\frac{\partial \ln Z}{\partial \varepsilon_r}\right) = \frac{1}{e^{\beta\varepsilon_r}-1} \qquad (15)$$

But, the energy and momentum of a photon are given by,

$$\varepsilon = h\nu \quad ; \quad p = \frac{\varepsilon}{c} = \frac{h\nu}{c} \qquad (16)$$

We assume the number of states associated with a given particle in a given region of phase space is proportional to the 6 dimensional volume of that region. The volume element in phase space is given by

$$d\Gamma = \iiint_V dxdydz \iiint_{p,p+dp} dp_x dp_y dp_z = V \iiint_{p,p+dp} dp_x dp_y dp_z = V(4\pi p^2 dp) \quad (17)$$

The volume in phase space for each discrete quantum state is h^3. So that phase space is divided into cells of volume h^3. Each cell represents one quantum state. We can see this by considering a small domain Γ_0 of the phase space in conjunction with Heisenberg's Uncertainty Principle.,

$$\Gamma_0 = \iiint_{\Gamma_0}\iiint dxdydzdp_x dp_y dp_z = \iiint_{\Gamma_0}\iiint (dxdp_x)(dydp_y)(dzdp_z) \quad (18)$$

$$\approx (\Delta x \Delta p_x)(\Delta y \Delta p_y)(\Delta z \Delta p_z) \geq \hbar.\hbar.\hbar = \hbar^3 \quad (19)$$

Therefore the minimum cell size $\approx h^3$. Therefore from (16) the number of discrete quantum states of a particle contained in a given volume V whose momentum lies in the range $p, p+dp$ is

$$f(p)dp = \frac{V 4\pi p^2 dp}{h^3} \quad (20)$$

In terms of the frequency, equation (16) gives for the number of photon states in which a photon has a frequency in the range $(v, v+dv)$

$$f(v)dv = \frac{8\pi V v^2 dv}{c^3} \quad (21)$$

There is an extra factor of 2 in (21), because a photon has an internal degree of freedom → polarization. A photon of a given momentum p can exist in 2 different states of polarization. From (15, 21) the number of photons in the frequency range $(v, v+dv)$ is given by

$$dN_v = <n> f(v)dv = \frac{8\pi V v^2 dv}{c^3(e^{\beta \varepsilon}-1)} \quad (22)$$

Therefore, the energy of the radiation in this frequency range is using (16)

$$dE_v dv = hv dN_v = \frac{8\pi V h v^3 dv}{c^3(e^{\beta \varepsilon}-1)} = \frac{8\pi V h v^3 dv}{c^3(e^{\beta hv}-1)} \quad (23)$$

3(c) If we integrate (23) over all frequencies v we obtain the total energy density (per unit volume V) of black body radiation at a given temperature T.

$$u(T) = \int_0^\infty \frac{8\pi h v^3}{c^3(e^{\beta hv}-1)} dv = \frac{8\pi h}{c^3}\left(\frac{k}{h}\right)^4 T^4 \int_0^\infty \frac{x^3}{(e^x-1)} dx \quad (24)$$

where,

$$x = \frac{hv}{kT} \tag{24a}$$

Since, $e^{-x} < 1$, for $0 < x < 1$, we have

$$\frac{1}{e^x - 1} = \frac{e^{-x}}{1 - e^{-x}} = \sum_{n=1}^{\infty} e^{-nx}$$

Let,

$$I = \int_0^{\infty} \frac{x^3}{e^x - 1} dx = \sum_{n=1}^{\infty} \int_0^{\infty} x^3 e^{-nx} dx \tag{24b}$$

Put, $y = nx$, then equation (24b) becomes,

$$I = \sum_{n=1}^{\infty} \frac{1}{n^4} \int_0^{\infty} y^3 e^{-y} dy \tag{24c}$$

By repeated integration by parts,

$$\int_0^{\infty} y^3 e^{-y} dy = \left[-y^3 e^{-y}\right]_0^{\infty} + \int_0^{\infty} 3y^2 e^{-y} dy = 3\int_0^{\infty} y^2 e^{-y} dy$$

$$= 3\left\{\left[-y^2 e^{-y}\right]_0^{\infty} + \int_0^{\infty} 2ye^{-y} dy\right\} = 6\int_0^{\infty} ye^{-y} dy$$

$$= 6\left\{\left[-ye^{-y}\right]_0^{\infty} + \int_0^{\infty} e^{-y} dy\right\} = 6 \tag{24d}$$

Substituting (24d) in (24c) gives,

$$I = 6\sum_{n=1}^{\infty} \frac{1}{n^4} \tag{24e}$$

Consider the Fourier Series for the function in the interval $[-\pi, \pi]$

$$f(x) = \frac{\pi}{4} \qquad 0 < x < \pi$$

$$= -\frac{\pi}{4} \qquad -\pi < x < 0 \tag{24f}$$

then,

$$f(x) = \frac{a_0}{2} + \sum_{n=1}^{\infty}(a_n \cos nx + b_n \sin nx) \tag{24g}$$

where,

$$a_n = \frac{1}{\pi}\int_{-\pi}^{\pi} f(x)\cos nx\, dx \qquad n \geq 0$$

$$b_n = \frac{1}{\pi}\int_{-\pi}^{\pi} f(x)\sin nx\, dx \qquad n \geq 1 \tag{24h}$$

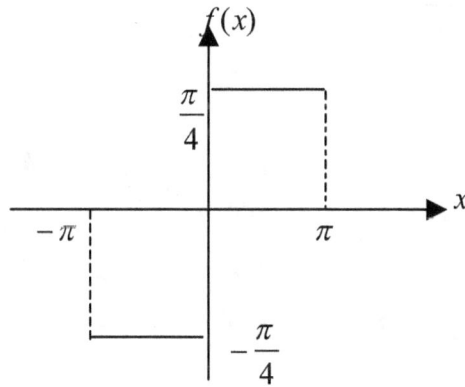

From (24f, 24h)

$$a_n = 0, \qquad n \geq 0$$

$$b_n = \frac{1}{2n}\left[1 - (-1)^n\right] \qquad n \geq 1 \tag{24i}$$

∴ From (24f-24i)

$$\frac{\pi}{4} = \sum_{n=1}^{\infty} \frac{1}{2n}\left[1 - (-1)^n\right]\sin nx \qquad \text{for} \quad 0 < x < \pi \tag{24j}$$

Therefore integrating from $(0, x)$,

$$\int_0^x \frac{\pi}{4}dx = -\sum_{n=1}^{\infty}\frac{1}{2n^2}[1-(-1)^n][\cos nx]_0^x \Rightarrow \frac{\pi x}{4} = \sum_{n=1}^{\infty}\frac{1}{2n^2}[1-(-1)^n](1-\cos nx) \tag{24k}$$

2 further integrations yield,

$$\frac{\pi x^2}{8} = \sum_{n=1}^{\infty} \frac{1}{2n^2}[1-(-1)^n]\left(x - \frac{\sin nx}{n}\right)$$

$$\frac{\pi x^3}{24} = \sum_{n=1}^{\infty} \frac{1}{2n^2}[1-(-1)^n]\left[\frac{x^2}{2} - \frac{1-\cos nx}{n^2}\right] \tag{24l}$$

Put, $x = \dfrac{\pi}{2}$ in (24k, 24l),

$$\frac{\pi^2}{8} = \sum_{m=0}^{\infty} \frac{1}{(2m+1)^2} \tag{24m}$$

$$\frac{\pi^4}{192} = \frac{\pi^2}{8} \sum_{m=0}^{\infty} \frac{1}{(2m+1)^2} - \sum_{m=0}^{\infty} \frac{1}{(2m+1)^4}$$

this latter equation gives using (24m)

$$\sum_{m=0}^{\infty} \frac{1}{(2m+1)^4} = \frac{\pi^4}{96} \tag{24n}$$

But,

$$\sum_{n=1}^{\infty} \frac{1}{n^4} = \sum_{m=0}^{\infty} \frac{1}{(2m+1)^4} + \sum_{m=1}^{\infty} \frac{1}{(2m)^4}$$

$$= \frac{\pi^4}{96} + \frac{1}{16}\sum_{n=1}^{\infty} \frac{1}{n^4}$$

This last step follows from (24n) and the fact that (m, n) are dummy suffices in the summation.

Therefore,

$$\sum_{n=1}^{\infty} \frac{1}{n^4} = \frac{\pi^4}{90} \tag{24o}$$

Substituting (24o) in (24e) gives,

$$I = \frac{\pi^4}{15} \tag{24p}$$

From (24b, 24p) equation (24) becomes,

$$u(T) = \frac{8\pi h}{c^3}\left(\frac{k}{h}\right)^4 T^4 I = \left(\frac{8\pi^5 k^4}{15 c^3 h^3}\right)T^4 \rightarrow \text{Stefan-Boltzmann law}$$

4. The Bohr model of the hydrogen atom → an electron of charge $-e$ and rest mass m_0 is moving in the field of a positively charged proton of charge $+e$ which is assumed to be at rest. The particle dynamics is described by the variational principle,

$$\delta \int_{t_1}^{t_2} L(r,\theta;\dot{r},\dot{\theta})\,dt \equiv \delta \int_{t_1}^{t_2} \left\{ -m_0 c^2 \sqrt{1 - \frac{\dot{r}^2 + r^2\dot{\theta}^2}{c^2}} + \frac{e^2}{r} \right\} dt = 0$$

Therefore, the Euler-Lagrange equations are,

$$\frac{d}{dt}\left(\frac{\partial L}{\partial \dot{r}}\right) - \frac{\partial L}{\partial r} = 0$$

$$\frac{d}{dt}\left(\frac{\partial L}{\partial \dot{\theta}}\right) - \frac{\partial L}{\partial \theta} = 0$$

This implies,

$$\frac{d}{dt}\left(\frac{m_0 \dot{r}}{\sqrt{1-\frac{v^2}{c^2}}}\right) - \frac{m_0 r\dot{\theta}^2}{\sqrt{1-\frac{v^2}{c^2}}} + \frac{e^2}{r^2} = 0 \quad \Rightarrow \quad \dot{p}_r = \frac{m_0 r\dot{\theta}^2}{\sqrt{1-\frac{v^2}{c^2}}} - \frac{e^2}{r^2}$$

$$\frac{d}{dt}\left(\frac{m_0 r^2 \dot{\theta}}{\sqrt{1-\frac{v^2}{c^2}}}\right) = 0 \quad \Rightarrow \quad p_\theta = \text{constant}$$

where,

$$v^2 = \dot{r}^2 + r^2\dot{\theta}^2 \tag{1}$$

and the generalized momenta are defined by,

$$p_r = \frac{\partial L}{\partial \dot{r}} = \frac{m_0 \dot{r}}{\sqrt{1-\frac{v^2}{c^2}}} \quad \text{(radial momentum)} \tag{2}$$

$$p_\theta = \frac{\partial L}{\partial \dot\theta} = \frac{m_0 r^2 \dot\theta}{\sqrt{1-\frac{v^2}{c^2}}} = \text{constant} \quad (\text{angular momentum}) \qquad (3)$$

Then the energy equation is given by

$$E = (m-m_0)c^2 - \frac{e^2}{r} = \frac{m_0 c^2}{\sqrt{1-\frac{v^2}{c^2}}} - m_0 c^2 - \frac{e^2}{r} \qquad (4)$$

where r is the distance between the electron and proton.

Applying the Sommerfeld quantization condition to the angular momentum

$$\oint p_\theta d\theta = n_\theta h \quad ; \quad n_\theta = 0, 1, 2, 3, \ldots \qquad (5)$$

Integrating,

$$\frac{m_0 r^2 \dot\theta}{\sqrt{1-\frac{v^2}{c^2}}} \cdot 2\pi = n_\theta h$$

$$\therefore \quad m_0 r^2 \dot\theta = n_\theta \hbar \sqrt{1-\frac{v^2}{c^2}} \qquad (6)$$

From (2) using (1)

$$p_r^2 = m_0^2 \frac{\dot r^2}{1-\frac{v^2}{c^2}} = m_0^2 \frac{v^2 - r^2 \dot\theta^2}{1-\frac{v^2}{c^2}} = m_0^2 c^2 \left(\frac{1}{1-\frac{v^2}{c^2}} - 1 \right) - \frac{m_0^2 r^2 \dot\theta^2}{1-\frac{v^2}{c^2}} \qquad (7)$$

Using (4)

$$\frac{1}{1-\frac{v^2}{c^2}} = \left(\frac{m_0 c^2 + \frac{e^2}{r} + E}{m_0 c^2} \right)^2 \qquad (8)$$

Using (6 & 8) in (7) gives,

$$p_r^2 = m_0^2 c^2 \left[\left(\frac{m_0 c^2 + \frac{e^2}{r} + E}{m_0 c^2} \right)^2 - 1 \right] - \left(\frac{n_\theta \hbar}{r} \right)^2 \qquad (9)$$

The radial momentum satisfies the quantum condition,

$$\oint p_r \, dr = n_r h \; ; \quad n_r = 0, 1, 2, 3, \ldots \qquad (10)$$

Substituting (9) in (10) gives on simplification,

$$\oint \sqrt{ \frac{1}{c^2} \left(E + \frac{e^2}{r} \right) \left(E + \frac{e^2}{r} + 2 m_0 c^2 \right) - \left(\frac{n_\theta \hbar}{r} \right)^2 } \, dr = n_r h$$

Further simplification of the integrand,

$$\oint \sqrt{ 2 m_0 E \left(1 + \frac{E}{2 m_0 c^2} \right) + 2 m_0 e^2 \left(1 + \frac{E}{m_0 c^2} \right) \frac{1}{r} - \frac{n_\theta^2 \hbar^2}{r^2} \left(1 - \frac{\alpha^2}{n_\theta^2} \right) } \, dr = n_r h \qquad (11)$$

where α is the fine structure constant defined by,

$$\alpha = \frac{e^2}{\hbar c} \qquad (12)$$

We know proceed to evaluate the integral,

$$I = \oint \sqrt{ -A + \frac{2B}{r} - \frac{C}{r^2} } \, dr = \oint \frac{\sqrt{ -Ar^2 + 2Br - C }}{r} \, dr \; , \quad \text{where} \quad A, B, C > 0$$

$$= \oint \frac{\sqrt{ -(Ar^2 - 2Br + C) }}{r} \, dr$$

If (λ, β) = roots of the quadratic, $Ar^2 - 2Br + C = 0$
then,

$$\lambda = \frac{B - \sqrt{B^2 - AC}}{A} \; ; \quad \beta = \frac{B + \sqrt{B^2 - AC}}{A} \; ; \quad \beta > \lambda$$

Put, $r = R + \frac{\lambda + \beta}{2} = R + \frac{B}{A}$

Therefore,

$$I = \oint \frac{\sqrt{-AR^2 + \frac{B^2 - AC}{A}}}{R + \frac{B}{A}} dR = \sqrt{A} \oint \frac{\sqrt{-R^2 + \frac{B^2 - AC}{A^2}}}{R + \frac{B}{A}} dR = \sqrt{A} \oint \frac{\sqrt{-R^2 + \rho^2}}{R + \sigma^2} dR$$

where, $\rho = \frac{\sqrt{B^2 - AC}}{A}$; $\sigma^2 = \frac{B}{A}$ \hfill (13)

The line integral implies going from $R = -\rho \to +\rho$ with a positive value for the square root in the above integrand, followed by $R = +\rho \to -\rho$, with a negative value for the square root resulting in

$$\therefore I = \sqrt{A} \int_{-\rho}^{\rho} \frac{\sqrt{-R^2 + \rho^2}}{R + \sigma^2} dR + \sqrt{A} \int_{+\rho}^{-\rho} -\frac{\sqrt{-R^2 + \rho^2}}{R + \sigma^2} dR = 2\sqrt{A} \int_{-\rho}^{+\rho} \frac{\sqrt{-R^2 + \rho^2}}{R + \sigma^2} dR$$

Put, $R = \rho \sin\theta \implies dR = \rho \cos\theta \, d\theta$

Therefore, $I = 2\rho^2 \sqrt{A} \int_{-\pi/2}^{+\pi/2} \frac{\cos^2\theta}{\rho \sin\theta + \sigma^2} d\theta = 2\sqrt{A} \int_{-\pi/2}^{\pi/2} \frac{\rho^2 (1 - \sin^2\theta)}{\rho \sin\theta + \sigma^2} d\theta$

$$= -2\sqrt{A} \int_{-\pi/2}^{\pi/2} \left[(\rho \sin\theta - \sigma^2) + \frac{\sigma^4 - \rho^2}{\rho \sin\theta + \sigma^2} \right] d\theta$$

$$= -2\sqrt{A} \left[-\sigma^2 \pi + \int_{-\pi/2}^{\pi/2} \frac{\sigma^4 - \rho^2}{\rho \sin\theta + \sigma^2} d\theta \right] = 2\sqrt{A} \left[\sigma^2 \pi + \int_{-\pi/2}^{\pi/2} \frac{\rho^2 - \sigma^4}{\rho \sin\theta + \sigma^2} d\theta \right]$$

Substituting the expressions for ρ, σ^2 from (13) in the above integral we have,

$$I = 2\sqrt{A} \left[\sigma^2 \pi - \frac{C}{A} \int_{-\pi/2}^{\pi/2} \frac{d\theta}{\rho \sin\theta + \sigma^2} \right]$$

$$= 2\sqrt{A} \left[\sigma^2 \pi - \frac{C}{A} \int_{-\pi/2}^{\pi/2} \frac{\left(\cos^2 \frac{\theta}{2} + \sin^2 \frac{\theta}{2}\right)}{2\rho \sin\frac{\theta}{2} \cos\frac{\theta}{2} + \sigma^2 \left(\sin^2 \frac{\theta}{2} + \cos^2 \frac{\theta}{2}\right)} d\theta \right]$$

Divide the numerator and the denominator of the integrand by $\cos^2 \frac{\theta}{2}$ we have

$$I = 2\sqrt{A}\left[\sigma^2\pi - \frac{C}{A}\int_{-\pi/2}^{\pi/2}\frac{\sec^2\frac{\theta}{2}}{\sigma^2\tan^2\frac{\theta}{2} + 2\rho\tan\frac{\theta}{2} + \sigma^2}d\theta\right]$$

Put, $\quad t = \tan\frac{\theta}{2} \Rightarrow dt = \frac{1}{2}\sec^2\frac{\theta}{2}$

$$\therefore I = 2\sqrt{A}\left[\sigma^2\pi - \frac{2C}{A\sigma^2}\int_{-1}^{1}\frac{dt}{t^2 + \frac{2\rho}{\sigma^2} + 1}\right]$$

$$= 2\sqrt{A}\left\{\sigma^2\pi - \frac{2C}{A\sigma^2}\left[\frac{1}{\sqrt{1 - \frac{\rho^2}{\sigma^4}}}\tan^{-1}\left(\frac{t + \frac{\rho}{\sigma^2}}{\sqrt{1 - \frac{\rho^2}{\sigma^4}}}\right)\right]_{-1}^{1}\right\}$$

$$= 2\sqrt{A}\left\{\sigma^2\pi - \frac{2C}{\sqrt{\sigma^4 - \rho^2}}\left[\tan^{-1}\left(\frac{1 + \frac{\rho}{\sigma^2}}{\sqrt{1 - \frac{\rho^2}{\sigma^4}}}\right) - \tan^{-1}\left(\frac{-1 + \frac{\rho}{\sigma^2}}{\sqrt{1 - \frac{\rho^2}{\sigma^4}}}\right)\right]\right\}$$

$$= 2\sqrt{A}\left\{\sigma^2\pi - \frac{2C}{A\sqrt{\frac{C}{A}}}\left[\tan^{-1}\left(\sqrt{\frac{1 + \frac{\rho}{\sigma^2}}{1 - \frac{\rho}{\sigma^2}}}\right) + \tan^{-1}\left(\sqrt{\frac{1 - \frac{\rho}{\sigma^2}}{1 + \frac{\rho}{\sigma^2}}}\right)\right]\right\}, \text{ using (13)}$$

$$= = 2\sqrt{A}\left(\sigma^2\pi - 2\sqrt{\frac{C}{A}}\frac{\pi}{2}\right), \text{ because for any } X\; ;\; \tan^{-1}X + \tan^{-1}\left(\frac{1}{X}\right) = \frac{\pi}{2}$$

Therefore, using (13)

$$\oint\sqrt{-A + \frac{2B}{r} - \frac{C}{r^2}}dr = 2\pi\left(\frac{B}{\sqrt{A}} - \sqrt{C}\right) \tag{14}$$

Using the result (14) in (11) gives,

$$2\pi\left\{\frac{m_0 e^2(1+\frac{E}{m_0 c^2})}{\sqrt{-2m_0 E(1+\frac{E}{2m_0 c^2})}} - n_\theta \hbar\sqrt{1-\frac{\alpha^2}{n_\theta^2}}\right\} = n_r h \tag{15}$$

Solving for E,

$$E = m_0 c^2\left\{-1 \pm \frac{1}{\sqrt{1+\frac{\alpha^2}{[n_r+\sqrt{n_\theta^2-\alpha^2}]^2}}}\right\}$$

5. By definition,

$$\{q_r, q_s\} = \sum_{k=1}^{k=n}\left(\frac{\partial q_r}{\partial q_k}\frac{\partial q_s}{\partial p_k} - \frac{\partial q_r}{\partial p_k}\frac{\partial q_s}{\partial q_k}\right) = 0$$

$$\{p_r, p_s\} = \sum_{k=1}^{k=n}\left(\frac{\partial p_r}{\partial q_k}\frac{\partial p_s}{\partial p_k} - \frac{\partial p_r}{\partial p_k}\frac{\partial p_s}{\partial q_k}\right) = 0$$

because,

$$\frac{\partial q_l}{\partial p_m} = \frac{\partial p_i}{\partial q_j} = 0 \quad \text{for} \quad i,j;l,m = 1 \to n$$

$$\{q_r, p_s\} = \sum_{k=1}^{k=n}\left(\frac{\partial q_r}{\partial q_k}\frac{\partial p_s}{\partial p_k} - \frac{\partial q_r}{\partial p_k}\frac{\partial p_s}{\partial q_k}\right) = \sum \delta_{rk}\delta_{sk} = \delta_{rs}$$

The commutator of 2 dynamical variables F, G is defined by,

$$[F,G] \equiv FG - GF$$

The transition to quantum mechanics is defined by,

$$[F,G] = i\hbar\{F,G\}$$

Therefore,

$$[q_r, q_s] = 0$$
$$[p_r, p_s] = 0$$
$$[q_r, p_s] = i\hbar \delta_{rs}$$

6(a). Dirac's wave equation is obtained by writing the relativistic Hamiltonian for a free particle

$$H = \pm c\sqrt{m_0^2 c^2 + \vec{p}^2} \qquad (1)$$

in the linearized form,

$$H = m_0 c^2 \alpha_0 + c\vec{\alpha} \cdot \vec{p} \qquad (2)$$

where, the operators $\alpha_0, \alpha_1, \alpha_2, \alpha_3$ satisfy the relations

$$\alpha_i \alpha_j + \alpha_j \alpha_i = 2\delta_{ij} \quad ; i,j = 0,1,2,3 \qquad (3)$$

In addition the $\{\alpha\}$ operators commute with the coordinate and momentum operators.

Proof: Squaring (2) using the summation convention gives,

$$H^2 = (m_0 c^2 \alpha_0 + c\alpha_l p_l)(m_0 c^2 \alpha_0 + c\alpha_m p_m) \qquad l,m = 1 \to 3$$

$$= m_0^2 c^4 \alpha_0^2 + m_0 c^3 (\alpha_0 \alpha_m p_m + \alpha_l p_l \alpha_0) + c^2 \alpha_l p_l \alpha_m p_m$$

$$= m_0^2 c^4 + m_0 c^3 (\alpha_0 \alpha_m p_m + \alpha_m p_m \alpha_0) + c^2 \alpha_l \alpha_m p_l p_m \qquad (4)$$

because, (l,m) are dummy suffices and $\alpha_0^2 = 1$ from (3)
Therefore,

$$\alpha_0 \alpha_m p_m + \alpha_m p_m \alpha_0 = (\alpha_0 \alpha_m + \alpha_m \alpha_0) p_m = 0 \qquad (5)$$

and,

$$\alpha_l \alpha_m p_l p_m = \alpha_m \alpha_l p_m p_l = \alpha_m \alpha_l p_l p_m = \frac{1}{2}(\alpha_l \alpha_m + \alpha_m \alpha_l) p_l p_m$$

$$= \delta_{lm} p_l p_m = p_l p_l = \vec{p}^2 \qquad (6)$$

Substituting (5,6) in (4) gives

$$H^2 = m_0^2 c^4 + c^2 \vec{p}^2 \quad \text{which is result (1)} \qquad \textbf{QED}$$

The operators $\alpha_0, \alpha_1, \alpha_2, \alpha_3$ can be represented by the 4 x 4 matrices

$$\alpha_0 = \begin{pmatrix} I_2 & 0_2 \\ 0_2 & -I_2 \end{pmatrix}; \; \alpha_1 = \begin{pmatrix} 0_2 & \sigma_x \\ \sigma_x & 0_2 \end{pmatrix}; \; \alpha_2 = \begin{pmatrix} 0_2 & \sigma_y \\ \sigma_y & 0_2 \end{pmatrix}; \; \alpha_3 = \begin{pmatrix} 0_2 & \sigma_z \\ \sigma_z & 0_2 \end{pmatrix} \qquad (7)$$

in which $I_2, 0_2$ are the 2 x 2 unit and zero matrices respectively, and $\sigma_x, \sigma_y, \sigma_z$ are the 2 x 2

Pauli spin matrices

$$\sigma_x = \begin{pmatrix} 0 & 1 \\ 1 & 0 \end{pmatrix}; \quad \sigma_y = \begin{pmatrix} 0 & -i \\ i & 0 \end{pmatrix}; \quad \sigma_z = \begin{pmatrix} 1 & 0 \\ 0 & -1 \end{pmatrix}; \tag{8}$$

(b) Define, the operators,
$$\pi_1 = -i\alpha_2\alpha_3 \ ; \quad \pi_2 = -i\alpha_3\alpha_1 \ ; \quad \pi_3 = -i\alpha_1\alpha_2 \tag{9}$$
Then a direct substitution shows,

$$\pi_1\pi_2 = -\pi_2\pi_1 = i\pi_3$$

$$\pi_2\pi_3 = -\pi_3\pi_2 = i\pi_1 \tag{10}$$

$$\pi_3\pi_1 = -\pi_1\pi_3 = i\pi_2$$

$$\pi_1^2 = \pi_2^2 = \pi_3^2 = 1$$

In addition, $\quad \vec{\pi} \wedge \vec{\pi} = 2i\vec{\pi} \tag{11}$

The eigenvalues of $(\pi_1, \pi_2, \pi_3) = \pm 1$
Define,
$$\vec{s} = \frac{\hbar}{2}\vec{\pi} \tag{12}$$
From (11, 12)

$$\vec{s} \wedge \vec{s} = i\hbar\vec{s} \tag{13}$$

By definition (13) corresponds to an angular momentum with eigenvalues $= \pm\dfrac{\hbar}{2}$

(c) It can be easily shown using (10) that for arbitrary three dimensional vectors $\vec{a}, \vec{b}$ whose components commute with (π_x, π_y, π_z),
$$(\vec{\pi}\bullet\vec{a})(\vec{\pi}\bullet\vec{b}) = \vec{a}\bullet\vec{b} + i\vec{\pi}\bullet(\vec{a}\wedge\vec{b}) \tag{14}$$

Proof: we use suffix notation, with a repeated index denoting summation. Then
Using the results (10)
$$(\vec{\pi}\bullet\vec{a})(\vec{\pi}\bullet\vec{b}) = (\pi_k a_k)(\pi_l b_l) = \pi_k\pi_l a_k b_l$$

$$= \sum_{k=l}^{3} \pi_k^2 a_k b_k + \sum_{k\neq l}^{3} \pi_k\pi_l a_k b_l$$

$$= \vec{a}\bullet\vec{b} + (\pi_1\pi_2 a_1 b_2 + \pi_1\pi_3 a_1 b_3 + \pi_2\pi_1 a_2 b_1 + \pi_2\pi_3 a_2 b_3 +$$
$$\pi_3\pi_1 a_3 b_1 + \pi_3\pi_2 a_3 b_2)$$

$$= \vec{a}\bullet\vec{b} + i(\pi_3 a_1 b_2 - \pi_2 a_1 b_3 - \pi_3 a_2 b_1 + \pi_1 a_2 b_3 + \pi_2 a_3 b_1 - \pi_1 a_3 b_2)$$

$$= \vec{a}\bullet\vec{b} + i[\pi_1(a_2 b_3 - a_3 b_2) + \pi_2(a_3 b_1 - a_1 b_3) + \pi_3(a_1 b_2 - a_2 b_1)]$$

$$= \vec{a}\bullet\vec{b} + i\vec{\pi}\bullet(\vec{a}\wedge\vec{b}) \qquad\qquad \textbf{QED}$$

Define,
$$\rho = -i\alpha_1\alpha_2\alpha_3 \tag{15}$$
From (10, 15),
$$\rho\pi_x = \alpha_1, \rho\pi_y = \alpha_2, \rho\pi_z = \alpha_3, \ \rho^2 = 1, \rho\alpha_0 = -\alpha_0\rho \tag{16}$$

In the presence of an electromagnetic field, for an electron of charge $-e$
$$H \to H + e\phi, \vec{p} \to \vec{p} + \frac{e}{c}\vec{A}$$

where $\phi, \vec{A}$ are respectively the scalar and vector potentials. Then from (2, 16)

$$H = c(m_0c\alpha_0 + \rho\vec{\pi}\bullet(\vec{p}+\frac{e}{c}\vec{A}) - e\phi$$

$$\therefore (H+e\phi)^2 = c^2\left\{m_0^2c^2 + \left[\vec{\pi}\bullet(\vec{p}+\frac{e}{c}\vec{A})\right]^2\right\} \tag{17}$$

From (14)

$$\left[\vec{\pi}\bullet(\vec{p}+\frac{e}{c}\vec{A})\right]^2 = \left(\vec{p}+\frac{e}{c}\vec{A}\right)^2 + i\vec{\pi}\bullet\left[\left(\vec{p}+\frac{e}{c}\vec{A}\right)\wedge\left(\vec{p}+\frac{e}{c}\vec{A}\right)\right]$$

But,

$$\left[\left(\vec{p}+\frac{e}{c}\vec{A}\right)\wedge\left(\vec{p}+\frac{e}{c}\vec{A}\right)\right] = -i\frac{e\hbar}{c}\vec{\nabla}\wedge\vec{A} = -i\frac{e\hbar}{c}\vec{B} \tag{18}$$

To prove result (18) consider the l th component of the vector product of the LHS

Proof: using suffix notation,
$$\left[\left(\vec{p}+\frac{e}{c}\vec{A}\right)\wedge\left(\vec{p}+\frac{e}{c}\vec{A}\right)\right]_l = \varepsilon_{ljk}\left(p_j+\frac{e}{c}A_j\right)\left(p_k+\frac{e}{c}A_k\right)$$

$$= \varepsilon_{ljk}\left[p_jp_k + \frac{e}{c}(A_jp_k+p_jA_k) + \frac{e^2}{c^2}A_jA_k\right]$$

$$= \frac{e}{c}\varepsilon_{ljk}(A_jp_k+p_jA_k) = \frac{e}{c}(\varepsilon_{lkj}A_kp_j + \varepsilon_{ljk}p_jA_k)$$

where, we have interchanged the dummy suffices (j,k) in the first term within the brackets.

ε_{ljk} is the permutation tensor defined by,

$$\varepsilon_{ljk} = +1, \text{ if } (l,j,k) \text{ is an even permutation of } (1,2,3)$$

$$= -1, \text{ if } (l,j,k) \text{ is an odd permutation of } (1,2,3)$$

$$= 0, \text{ if any 2 of the indices } (l,j,k) \text{ are equal}$$

Therefore,

$$\left[\left(\vec{p}+\frac{e}{c}\vec{A}\right)\wedge\left(\vec{p}+\frac{e}{c}\vec{A}\right)\right]_l = \frac{e}{c}(-\varepsilon_{ljk}A_k p_j + \varepsilon_{ljk}p_j A_k)$$

$$= \frac{e}{c}\varepsilon_{ljk}(-A_k p_j + p_j A_k)$$

$$= -i\frac{\hbar e}{c}\varepsilon_{ljk}(-A_k\frac{\partial}{\partial x_j} + \frac{\partial}{\partial x_j}A_k) \quad (\because \vec{p} = -i\hbar\vec{\nabla})$$

$$= -i\frac{\hbar e}{c}\varepsilon_{ljk}\frac{\partial A_k}{\partial x_j} = -i\frac{\hbar e}{c}(\vec{\nabla}\wedge\vec{A})_l$$

Therefore, $\left[\left(\vec{p}+\frac{e}{c}\vec{A}\right)\wedge\left(\vec{p}+\frac{e}{c}\vec{A}\right)\right] = -i\frac{\hbar e}{c}\vec{\nabla}\wedge\vec{A} = -i\frac{\hbar e}{c}\vec{B}$ **QED**

$$(H+e\phi)^2 = c^2(m_0^2 c^2 + |\vec{p}+\frac{e}{c}\vec{A}|^2 + \frac{e\hbar}{c}\vec{\pi}\bullet\vec{B}) \tag{19}$$

Extra term in (19) = $\frac{e\hbar}{c}\vec{\pi}\bullet\vec{B}$

7. In Maxwell's Equations the current vector **j** is defined by
$$\mathbf{j} = \rho\mathbf{v}$$
where ρ is the charge density and **v** is the velocity vector.

Maxwell's Equations in a vacuum :

$$\nabla\bullet\mathbf{E} = 4\pi\rho \;;\; \nabla\bullet\mathbf{H} = 0 \;;\; \nabla\wedge\mathbf{E} = -\frac{1}{c}\frac{\partial}{\partial t}\mathbf{H} \;;\; \nabla\wedge\mathbf{H} = \frac{4\pi}{c}\mathbf{j} + \frac{1}{c}\frac{\partial}{\partial t}\mathbf{E}$$

$$\nabla\bullet\mathbf{H} = 0 \Rightarrow \mathbf{H} = \nabla\wedge\mathbf{A}$$

$$\therefore \nabla\wedge\mathbf{E} = -\frac{1}{c}\frac{\partial}{\partial t}\nabla\wedge\mathbf{A} \Rightarrow \nabla\wedge(\mathbf{E} + \frac{1}{c}\frac{\partial}{\partial t}\mathbf{A}) = 0$$

∴ ∃ ϕ such that,

$$\mathbf{E} + \frac{1}{c}\frac{\partial}{\partial t}\mathbf{A} = -\nabla\phi \Rightarrow \mathbf{E} = -\nabla\phi - \frac{1}{c}\frac{\partial}{\partial t}\mathbf{A}$$

Again,

$$\nabla \wedge (\nabla \wedge \mathbf{A}) = \frac{4\pi}{c}\mathbf{j} + \frac{1}{c}\frac{\partial}{\partial t}\left(-\nabla\phi - \frac{1}{c}\frac{\partial}{\partial t}\mathbf{A}\right)$$

i.e.

$$\nabla(\nabla \cdot \mathbf{A}) - \nabla^2 \mathbf{A} = \frac{4\pi}{c}\mathbf{j} - \frac{1}{c^2}\frac{\partial^2}{\partial t^2}\mathbf{A} - \frac{1}{c}\nabla\left(\frac{\partial\phi}{\partial t}\right)$$

Simplifying, $\quad \nabla^2 \mathbf{A} - \frac{1}{c^2}\frac{\partial^2}{\partial t^2}\mathbf{A} = -\frac{4\pi}{c}\mathbf{j} + \nabla(\nabla \cdot \mathbf{A} + \frac{1}{c}\frac{\partial\phi}{\partial t})$

Gauge Condition: $\quad \nabla \cdot \mathbf{A} + \frac{1}{c}\frac{\partial\phi}{\partial t} = 0$

Therefore,

$$\nabla^2 \mathbf{A} - \frac{1}{c^2}\frac{\partial^2}{\partial t^2}\mathbf{A} = -\frac{4\pi}{c}\mathbf{j}$$

Similarly using the Gauge Condition,

$$\nabla^2 \phi - \frac{1}{c^2}\frac{\partial^2}{\partial t^2}\phi = -4\pi\rho$$

Proof: for any arbitrary well behaved functions of position and time

$$\nabla^2\left(\frac{f(\vec{r},t)}{g(\vec{r},t)}\right) = f\nabla^2\left(\frac{1}{g}\right) + \frac{1}{g}\nabla^2 f + 2\nabla\left(\frac{1}{g}\right) \cdot \nabla f$$

Take, $g = \frac{1}{|\vec{r}-\vec{r}'|} \Rightarrow \nabla^2 g = -4\pi\delta(\vec{r}-\vec{r}')$

$$f = \rho(\vec{r}', t - \frac{|\vec{r}-\vec{r}'|}{c})$$

Consider the integral,

$$\phi(\vec{r},t) = \int_V \frac{\rho(\vec{r}', t - \frac{|\vec{r}-\vec{r}'|}{c})}{|\vec{r}-\vec{r}'|} dV' \qquad (1)$$

Then,

$$\frac{\partial^2}{\partial t^2}\phi(\vec{r},t) = \frac{\partial^2}{\partial t^2}\int_V \frac{\rho(\vec{r}',t-\frac{|\vec{r}-\vec{r}'|}{c})}{|\vec{r}-\vec{r}'|}dV' = \int_V \frac{1}{|\vec{r}-\vec{r}'|}\frac{\partial^2}{\partial T^2}\rho(\vec{r}',T)dV' \qquad (2)$$

where,

$$T = t - \frac{|\vec{r}-\vec{r}'|}{c} \; ; \quad \vec{r} = \vec{r}(t) \qquad (3)$$

and,

$$\nabla^2\phi(\vec{r},t) = \nabla^2\int_V \frac{\rho(\vec{r}',t-\frac{|\vec{r}-\vec{r}'|}{c})}{|\vec{r}-\vec{r}'|}dV'$$

$$= \int_V \left[\rho(\vec{r}',t-\frac{|\vec{r}-\vec{r}'|}{c})\nabla_r^2\left(\frac{1}{|\vec{r}-\vec{r}'|}\right) + \frac{1}{|\vec{r}-\vec{r}'|}\nabla_r^2\rho(\vec{r}',t-\frac{|\vec{r}-\vec{r}'|}{c}) + 2\nabla_r\left(\frac{1}{|\vec{r}-\vec{r}'|}\right)\bullet\nabla_r\rho(\vec{r}',t-\frac{|\vec{r}-\vec{r}'|}{c})\right]$$

$$= -4\pi\rho(\vec{r},t) + \int_V \left[\frac{1}{|\vec{r}-\vec{r}'|}\nabla_r^2\rho(\vec{r}',T) + 2\nabla_r\left(\frac{1}{|\vec{r}-\vec{r}'|}\right)\bullet\nabla_r\rho(\vec{r}',T)\right]dV' \qquad (4)$$

But,

$$\nabla_r^2\rho(\vec{r}',T) = \frac{\partial^2\rho}{\partial x_k \partial x_k} \text{ , using the summation convention} \rightarrow \text{ sum over repeated index}$$

$$= \frac{\partial}{\partial x_k}\left(\frac{\partial\rho}{\partial x_k}\right) = \frac{\partial}{\partial x_k}\left(\frac{\partial\rho}{\partial T}\cdot\frac{\partial T}{\partial x_k}\right) = \frac{\partial\rho}{\partial T}\frac{\partial}{\partial x_k}\left(\frac{\partial T}{\partial x_k}\right) + \frac{\partial T}{\partial x_k}\cdot\frac{\partial}{\partial x_k}\left(\frac{\partial\rho}{\partial T}\right)$$

and,

$$\frac{\partial}{\partial x_k} = \frac{\partial T}{\partial x_k}\frac{\partial}{\partial T}$$

$$\therefore \nabla_r^2\rho(\vec{r}',T) = \frac{\partial\rho}{\partial T}\nabla_r^2 T + \frac{\partial^2\rho}{\partial T^2}\nabla_r T \bullet \nabla_r T \qquad (5)$$

and,

$$\nabla_r\rho(\vec{r}',T) = \frac{\partial\rho}{\partial T}\nabla_r T \qquad (6)$$

$$\nabla_r\left(\frac{1}{|\vec{r}-\vec{r}'|}\right) = -\frac{\vec{r}-\vec{r}'}{|\vec{r}-\vec{r}'|^3} \qquad (7)$$

$$\nabla_r T = -\frac{1}{c}\frac{\vec{r}-\vec{r}'}{|\vec{r}-\vec{r}'|} \qquad (8)$$

$$\nabla_r^2 T = \nabla_r \bullet \nabla_r T = -\frac{1}{c}\nabla_r \bullet \left(\frac{\vec{r}-\vec{r}'}{|\vec{r}-\vec{r}'|}\right) = -\frac{1}{c}\left(\frac{3}{|\vec{r}-\vec{r}'|} - \frac{1}{|\vec{r}-\vec{r}'|}\right) = -\frac{2}{c|\vec{r}-\vec{r}'|} \qquad (9)$$

Substituting (5 -9) in (4) and simplifying,

$$\nabla^2 \phi(\vec{r},t) = -4\pi\rho(\vec{r},t) + \frac{1}{c^2}\int\frac{1}{|\vec{r}-\vec{r}'|}\frac{\partial^2 \rho}{\partial T^2}dV'$$

From (2),

$$\nabla^2 \phi - \frac{1}{c^2}\frac{\partial^2 \phi}{\partial t^2} = -4\pi\rho(\vec{r},t)$$

A similar result holds for the vector potential $\vec{A}(\vec{r},t)$

8. **Biot- Savart Law** : From exercise 5 we see that for static fields, $\frac{\partial}{\partial t} = 0$

$$\therefore \nabla^2 \mathbf{A} = -\frac{4\pi}{c}\mathbf{j} \quad ; \quad \nabla^2 \phi = -4\pi\rho$$

$$\therefore \mathbf{A}(\mathbf{r}) = \frac{1}{c}\int_V \frac{\vec{j}(\vec{r}')}{|\vec{r}-\vec{r}'|}dV' \quad ;\text{because } \nabla^2\left(\frac{1}{|\vec{r}-\vec{r}'|}\right) = -4\pi\delta(\vec{r}-\vec{r}')$$

$$\therefore \mathbf{H} = \frac{1}{c}\int_V \nabla_r \wedge \left(\frac{\vec{j}(\vec{r}')}{|\vec{r}-\vec{r}'|}\right)dV' = \frac{1}{c}\int_V \nabla_r\left(\frac{1}{|\vec{r}-\vec{r}'|}\right) \wedge \vec{j}(\vec{r}')dV'$$

$$= -\frac{1}{c}\int_V \frac{(\vec{r}-\vec{r}') \wedge \vec{j}(\vec{r}')}{|\vec{r}-\vec{r}'|^3}dV'$$

But, $\int_V \mathbf{j}(\mathbf{r}')dV' = \int_V \rho\, \mathbf{v}(\mathbf{r}')\,dV' = \int_V dq\, \mathbf{v}(\mathbf{r}') = \int_V dt\frac{dq}{dt}\mathbf{v}(\mathbf{r}') = \int_V dt.i\,\mathbf{v}(\mathbf{r}') = \int i d\mathbf{s}$

where,

$dq = \rho(\vec{r}')dV'$ = total charge in a volume element dV'

$i = \frac{dq}{dt}$ = current ; $d\mathbf{s} = \mathbf{v}(\mathbf{r}')\,dt$ = element of length

$\therefore \vec{j}(\vec{r}')dV' = id\vec{s}'$

$$\therefore \mathbf{H} = -\frac{1}{c}\int_L i\frac{(\vec{r}-\vec{r}')\wedge d\vec{s}'}{|\vec{r}-\vec{r}'|^3} = \frac{1}{c}\int_L i\frac{d\vec{s}'\wedge(\vec{r}-\vec{r}')}{|\vec{r}-\vec{r}'|^3}$$

This implies

$$d\mathbf{H} = \frac{i}{c}\frac{d\vec{s}'\wedge(\vec{r}-\vec{r}')}{|\vec{r}-\vec{r}'|^3}$$

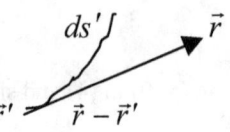

The magnetic field at a field point r_2 due to a current element $i_1 ds_1$ at a source point r_1 is given by

$$d\mathbf{H}_{21} = \frac{i_1}{c}\frac{d\vec{s}_1 \wedge (\vec{r}_2-\vec{r}_1)}{|\vec{r}_2-\vec{r}_1|^3}$$

The force experienced by a charge $\rho_2 dV_2$ at r_2 moving with a velocity v_2 is

$$d\mathbf{F}_{21} = \frac{\rho_2 dV_2}{c}\vec{v}_2 \wedge d\mathbf{H}_{21} = \frac{\vec{j}_2 dV_2 \wedge d\vec{H}_{21}}{c} = \frac{i_2 d\vec{s}_2 \wedge d\vec{H}_{21}}{c}$$

Substituting for $d\mathbf{H}_{21}$ and integrating gives the required result

9. Let a particle of charge q be at position $\vec{r}'(T)$ at time T moving at a velocity $\vec{v}'(T)$. Then,

$$\rho(\vec{r},T) = q\delta(\vec{r}-\vec{r}'(T))$$

$$\therefore \rho(\vec{r}',T) = q\delta(\vec{r}'-\vec{r}'(T)) \qquad (1)$$

From Exercise 7 and equation (1),

$$\phi(\vec{r},t) = \int_{allspace} \frac{\rho\left(\vec{r}',t\pm\frac{|\vec{r}-\vec{r}'|}{c}\right)}{|\vec{r}-\vec{r}'|}dV' \qquad (2)$$

$$= \int_{allspace} \frac{q\rho(\vec{r}',T)\delta\left(T-(t\pm\frac{|\vec{r}-\vec{r}'|}{c})\right)}{|\vec{r}-\vec{r}'|}dTdV'$$

$$= q\int_{alltime} \frac{\delta\left(T-(t\pm\frac{|\vec{r}-\vec{r}'(T)|}{c})\right)}{|\vec{r}-\vec{r}'(T)|}dT \qquad (3)$$

Put,

$$\xi = T - t \mp \frac{|\vec{r} - \vec{r}'(T)|}{c} \tag{4}$$

$$\therefore \quad \frac{\partial \xi}{\partial T} = 1 \pm \frac{(\vec{r} - \vec{r}'(T)) \bullet \vec{v}'(T)}{c|\vec{r} - \vec{r}'(T)|} = 1 \pm \frac{\vec{\varepsilon}'(T) \bullet \vec{v}'(T)}{c} \tag{5}$$

Substituting (4 & 5) in (3) gives,

$$\phi(\vec{r},t) = q \int \frac{\delta(\xi)}{|\vec{r} - \vec{r}'(T)|[1 \pm \frac{\vec{\varepsilon}'(T) \bullet \vec{v}'(T)}{c}]} d\xi$$

$$= \frac{q}{c|\vec{r} - \vec{r}'(T)|\left[1 \pm \frac{\vec{\varepsilon}'(T) \bullet \vec{v}'(T)}{c}\right]}$$

where, $\xi = 0 \rightarrow T = t \pm \frac{|\vec{r} - \vec{r}'(T)|}{c}$

A similar result can be obtained for the vector potential $\vec{A}(\vec{r},t)$.

10. Lienard-Wiechert Potentials

$$T = t - \frac{|\vec{r} - \vec{r}'(T)|}{c} \tag{1}$$

$$\therefore \quad \frac{\partial T}{\partial t} = 1 - \frac{1}{c} \frac{\partial |\vec{r} - \vec{r}'(T)|}{\partial T} \frac{\partial T}{\partial t} \tag{2}$$

But, $\dfrac{\partial}{\partial T}|\vec{r} - \vec{r}'(T)| = -\dfrac{[\vec{r} - \vec{r}'(T)] \bullet \vec{v}'(T)}{|\vec{r} - \vec{r}'(T)|} = -\vec{\varepsilon}' \bullet \vec{v}' \tag{3}$

where,

$$\vec{\varepsilon}'(T) = \frac{\vec{r} - \vec{r}'(T)}{|\vec{r} - \vec{r}'(T)|} \quad ; \quad \vec{v}'(T) = \frac{d\vec{r}'(T)}{dT} \tag{4}$$

Substituting (3) in (2)

$$\frac{\partial T}{\partial t} = \frac{1}{1 - \dfrac{\vec{v}' \bullet \vec{\varepsilon}'}{c}} \tag{5}$$

Consider,

$$\frac{\partial}{\partial x_1}|\vec{r}-\vec{r}'(T)| = \frac{\partial}{\partial x_1}\sqrt{[x_1-x_1'(T)]^2+[x_2-x_2'(T)]^2+[x_3-x_3'(T)]^2} \quad ; \quad T=T(x_1,x_2,x_3;t)$$

$$=\frac{1}{|\vec{r}-\vec{r}'(T)|}\left\{[x_1-x_1'(T)][1-\frac{dx_1'(T)}{dT}\frac{\partial T}{\partial x_1}]-[x_2-x_2'(T)]\frac{dx_2'(T)}{dT}\frac{\partial T}{\partial x_1}-[x_3-x_3'(T)]\frac{dx_3'(T)}{dT}\frac{\partial T}{\partial x_1}\right\}$$

$$=\frac{1}{|\vec{r}-\vec{r}'|}\left\{[x_1-x_1'(T)]-\frac{\partial T}{\partial x_1}[\vec{r}-\vec{r}'(T)]\bullet\frac{d\vec{r}'(T)}{dT}\right\}$$

Similar expressions hold for the partial derivatives with respect to x_2, x_3.

Therefore combining these expressions into a single vector equation we have,,

$$\nabla_r|\vec{r}-\vec{r}'| = \vec{\varepsilon}'(T) - \vec{\varepsilon}'(T)\bullet\vec{v}'(T)\nabla_r T$$

$$\therefore \quad \nabla_r T = -\frac{1}{c}\nabla_r|\vec{r}-\vec{r}'| = -\frac{1}{c}\left[\vec{\varepsilon}'(T) - \vec{\varepsilon}'(T)\bullet\vec{v}'(T)\nabla_r T\right]$$

$$\therefore \quad \nabla_r T = -\frac{\vec{\varepsilon}'(T)}{c\left[1-\dfrac{\vec{\varepsilon}'(T)\bullet\vec{v}'(T)}{c}\right]} \tag{6}$$

We include the following results for ready reference,

$$\frac{\partial \vec{\varepsilon}'(T)}{\partial T} = \frac{\partial}{\partial T}\left(\frac{\vec{r}-\vec{r}'}{|\vec{r}-\vec{r}'|}\right) = \frac{(\vec{r}-\vec{r}')\bullet\vec{v}'(T)}{|\vec{r}-\vec{r}'|^2}(\vec{r}-\vec{r}') - \frac{\vec{v}'(T)}{|\vec{r}-\vec{r}'|} = \frac{\vec{\varepsilon}'(T)\bullet\vec{v}'(T)}{|\vec{r}-\vec{r}'|} - \frac{\vec{v}'(T)}{|\vec{r}-\vec{r}'|}$$

$$= \frac{\vec{\varepsilon}'\wedge(\vec{\varepsilon}'\wedge\vec{v}')}{|\vec{r}-\vec{r}'|} \tag{7}$$

and,

$$\frac{\partial}{\partial T}\left[|\vec{r}-\vec{r}'|\left(1-\frac{\vec{v}'\bullet\vec{\varepsilon}'}{c}\right)\right] = -\frac{1}{c}|\vec{r}-\vec{r}'|\left[(\vec{a}'\bullet\vec{\varepsilon}') + \vec{v}'\bullet\frac{\partial\vec{\varepsilon}'}{\partial T}\right] + \left(1-\frac{\vec{v}'\bullet\vec{\varepsilon}'}{c}\right)\frac{\partial|\vec{r}-\vec{r}'|}{\partial T}$$

where, $\quad \vec{a}'(T) = \dfrac{d\vec{v}'(T)}{dT}$

using equations (3, 7)

$$\frac{\partial}{\partial T}\left[|\vec{r}-\vec{r}'|\left(1-\frac{\vec{v}'\bullet\vec{\varepsilon}'}{c}\right)\right] = -\frac{1}{c}|\vec{r}-\vec{r}'|\left[(\vec{a}'\bullet\vec{\varepsilon}') + \vec{v}'\bullet\frac{\vec{\varepsilon}'\wedge(\vec{\varepsilon}'\wedge\vec{v}')}{|\vec{r}-\vec{r}'|}\right] - \left(1-\frac{\vec{v}'\bullet\vec{\varepsilon}'}{c}\right)\vec{v}'\bullet\vec{\varepsilon}'$$

$$= -\frac{1}{c}|\vec{r}-\vec{r}'|(\vec{a}'\bullet\vec{\varepsilon}') - \vec{v}'\bullet\left(\vec{\varepsilon}' - \frac{\vec{v}'}{c}\right) \qquad (8)$$

Consider, $\quad \nabla_r[\vec{v}'(T)\bullet\vec{\varepsilon}'(T)]\Big|_{T=\text{constant}} = (\vec{v}'\bullet\nabla_r)\vec{\varepsilon}' + \vec{v}'\wedge(\nabla_r\wedge\vec{\varepsilon}') \qquad (9)$

where we have used the vector identity,

$$\nabla(\vec{a}\bullet\vec{b}) = (\vec{a}\bullet\nabla)\vec{b} + (\vec{b}\bullet\nabla)\vec{a} + \vec{a}\wedge(\nabla\wedge\vec{b}) + \vec{b}\wedge(\nabla\wedge\vec{a})$$

using suffix notation, keeping *T= constant*

$$(\vec{v}'\bullet\nabla_r)\vec{\varepsilon}' = v'_j\frac{\partial}{\partial X_j}\left(\frac{X_i}{R}\right) = v'_j\left(\frac{1}{R}\delta_{ij} - \frac{X_iX_j}{R^3}\right) = \frac{v'_i}{R} - \frac{\vec{v}'\bullet\vec{\varepsilon}'}{R^2} = \frac{v'_i - (\vec{v}'\bullet\vec{\varepsilon}')\varepsilon'_i}{R} \qquad (10)$$

where, $X_j = x_j - x'_j(T)$; $\varepsilon'_j = \frac{X_j}{R}$; j =1,2,3 and $R = \sqrt{X_1^2 + X_2^2 + X_3^2}$

using (9 & 10)

$$\nabla_r\phi\big|_{T=\text{constant}} = \nabla_r\left[\frac{q}{|\vec{r}-\vec{r}'|\left(1-\frac{\vec{v}'(T)\bullet\vec{\varepsilon}'(T)}{c}\right)}\right] = -\frac{q(\vec{\varepsilon}'(T) - \vec{v}'(T)/c)}{|\vec{r}-\vec{r}'|^2\left(1-\frac{\vec{v}'(T)\bullet\vec{\varepsilon}'(T)}{c}\right)^2} \qquad (11)$$

Again using (8),

$$\frac{\partial\phi}{\partial T} = -\frac{q}{|\vec{r}-\vec{r}'|^2\left(1-\frac{\vec{v}'(T)\bullet\vec{\varepsilon}'(T)}{c}\right)^2}\frac{\partial}{\partial T}\left\{|\vec{r}-\vec{r}'|\left(1-\frac{\vec{v}'(T)\bullet\vec{\varepsilon}'(T)}{c}\right)\right\}$$

$$= -\frac{q}{|\vec{r}-\vec{r}'|^2\left(1-\frac{\vec{v}'\bullet\vec{\varepsilon}'}{c}\right)^2}\left[-\frac{|\vec{r}-\vec{r}'|}{c}(\vec{a}'\bullet\vec{\varepsilon}') - \vec{v}'\bullet(\vec{\varepsilon}' - \frac{\vec{v}'}{c})\right] \qquad (12)$$

Finally, from (6, 11, 12)

$$\nabla_r\phi = \nabla_r\phi\big|_{T=\text{constant}} + \frac{\partial\phi}{\partial T}\nabla_r T$$

$$= -q \frac{\left[\vec{\varepsilon}'\left(1-\frac{\vec{v}'^2}{c^2}\right) - \frac{\vec{v}'}{c}\left(1-\frac{\vec{v}'\bullet\vec{\varepsilon}'}{c}\right)\right]}{|\vec{r}-\vec{r}'|^2 \left(1-\frac{\vec{v}'\bullet\vec{\varepsilon}'}{c}\right)^3} - \frac{q\vec{\varepsilon}'(\vec{a}\bullet\vec{\varepsilon}')}{c^2|\vec{r}-\vec{r}'|\left(1-\frac{\vec{v}'\bullet\vec{\varepsilon}'}{c}\right)^3} \qquad (13)$$

From (3,7) the Lienard-Wiechert vector potential gives,

$$\frac{\partial \vec{A}(\vec{r},t)}{\partial t} = \frac{\partial \vec{A}(\vec{r},t)}{\partial T}\frac{\partial T}{\partial t} = \frac{\partial}{\partial T}\left(\frac{q\vec{v}'(T)}{c|\vec{r}-\vec{r}'|\left[1-\frac{\vec{v}'\bullet\vec{\varepsilon}'}{c}\right]}\right)\frac{\partial T}{\partial t}$$

But,

$$\frac{\partial \vec{A}}{\partial T} = \frac{q\vec{a}'}{c|\vec{r}-\vec{r}'|\left(1-\frac{\vec{v}'\bullet\vec{\varepsilon}'}{c}\right)} + \frac{q\vec{v}'}{c|\vec{r}-\vec{r}'|^2\left(1-\frac{\vec{v}'\bullet\vec{\varepsilon}'}{c}\right)^2}\left[\frac{|\vec{r}-\vec{r}'|}{c}(\vec{a}'\bullet\vec{\varepsilon}') + \vec{v}'\bullet\left(\vec{\varepsilon}'-\frac{\vec{v}'}{c}\right)\right] \qquad (13a)$$

∴ From (5)

$$-\frac{1}{c}\frac{\partial \vec{A}}{\partial t} = -\frac{q\vec{a}'}{c^2|\vec{r}-\vec{r}'|\left(1-\frac{\vec{v}'\bullet\vec{\varepsilon}'}{c}\right)^2} - \frac{q\frac{\vec{v}'}{c}(\vec{a}'\bullet\vec{\varepsilon}')}{c^2|\vec{r}-\vec{r}'|\left(1-\frac{\vec{v}'\bullet\vec{\varepsilon}'}{c}\right)^3}$$

$$-\frac{q\frac{\vec{v}'}{c}\left[\frac{\vec{v}'}{c}\bullet\left(\vec{\varepsilon}'-\frac{\vec{v}'}{c}\right)\right]}{|\vec{r}-\vec{r}'|^2\left(1-\frac{\vec{v}'\bullet\vec{\varepsilon}'}{c}\right)^3}$$

$$= -\frac{q\frac{\vec{v}'}{c}\left[\frac{\vec{v}'}{c}\bullet\left(\vec{\varepsilon}'-\frac{\vec{v}'}{c}\right)\right]}{|\vec{r}-\vec{r}'|^2\left(1-\frac{\vec{v}'\bullet\vec{\varepsilon}'}{c}\right)^3} + \frac{-q\vec{a}' + q\vec{\varepsilon}'\wedge(\vec{a}'\wedge\frac{\vec{v}'}{c})}{c^2|\vec{r}-\vec{r}'|\left(1-\frac{\vec{v}'\bullet\vec{\varepsilon}'}{c}\right)^3} \qquad (14)$$

From equations (13, 14) the electric field is given by,

$$\vec{E}(\vec{r},t) = -\nabla_r \phi(\vec{r},t) - \frac{1}{c}\frac{\partial \vec{A}(\vec{r},t)}{\partial t}$$

$$= \frac{-q[\vec{a}' - (\vec{a}' \bullet \vec{\varepsilon}')\vec{\varepsilon}'] + q\vec{\varepsilon}' \wedge \left(\vec{a}' \wedge \frac{\vec{v}'}{c}\right)}{c^2 |\vec{r} - \vec{r}'| \left(1 - \frac{\vec{v}' \bullet \vec{\varepsilon}'}{c}\right)^3} + \frac{q\left(\vec{\varepsilon}' - \frac{\vec{v}'}{c}\right)\left(1 - \frac{v'^2}{c^2}\right)}{|\vec{r} - \vec{r}'|^2 \left(1 - \frac{\vec{v}' \bullet \vec{\varepsilon}'}{c}\right)^3} \tag{15}$$

To compute the magnetic field let us first consider the *i*-*t*h component of Curl $\vec{A}$ using suffix notation,

$$\left(\nabla_r \wedge \vec{A}(\vec{r},t)\right)_i = e_{ijk} \frac{\partial A_k}{\partial x_j} = e_{ijk} \left[\left(\frac{\partial A_k}{\partial x_j}\right)_{T=\text{constant}} + \frac{\partial A_k}{\partial T} \frac{\partial T}{\partial x_j} \right] \tag{16}$$

where e_{ijk} is the permutation tensor defined by,

$$e_{ijk} = +1 \quad \text{if } (i, j, k) \text{ is an even permutation of } (1, 2, 3)$$

$$= -1 \quad \text{if } (i, j, k) \text{ is an odd permutation of } (1, 2, 3)$$

$$= 0 \quad \text{if any 2 or more of } (i, j, k) \text{ are equal}$$

Therefore from (16),

$$\nabla_r \wedge \vec{A} = \left(\nabla_r \wedge \vec{A}\right)_{T=\text{constant}} - \frac{\partial \vec{A}}{\partial T} \wedge \nabla_r T \qquad (\because e_{ijk} = -e_{ikj}) \tag{17}$$

But,

$$\left(\nabla_r \wedge \vec{A}\right)_{T=\text{constant}} = \left[\nabla_r \wedge \left(\frac{q\vec{v}'(T)}{c|\vec{r} - \vec{r}'(T)|(1 - \frac{\vec{v}'(T) \bullet \vec{\varepsilon}'(T)}{c})}\right)\right]_{T=\text{constant}} \tag{18}$$

using the vector identity,

$$\nabla \wedge (\lambda \vec{F}) = \lambda (\nabla \wedge \vec{F}) + \nabla \lambda \wedge \vec{F} \tag{19}$$

in (18) we have in conjunction with (11)

$$\left(\nabla_r \wedge \vec{A}\right)_{T=\text{constant}} = -\frac{q\left(\vec{\varepsilon}' - \frac{\vec{v}'}{c}\right) \wedge \frac{\vec{v}'}{c}}{|\vec{r}-\vec{r}'|^2 \left(1-\frac{\vec{v}' \bullet \vec{\varepsilon}'}{c}\right)^2} \qquad (20)$$

From (6, 13a)

$$-\frac{\partial \vec{A}}{\partial T} = \frac{-q\vec{a}' + q\vec{\varepsilon}' \wedge \left(\vec{a}' \wedge \frac{\vec{v}'}{c}\right)}{c|\vec{r}-\vec{r}'|\left(1-\frac{\vec{v}' \bullet \vec{\varepsilon}'}{c}\right)^2} - \frac{q\vec{v}'\left[\frac{\vec{v}'}{c} \bullet \left(\vec{\varepsilon}' - \frac{\vec{v}'}{c}\right)\right]}{|\vec{r}-\vec{r}'|^2 \left(1-\frac{\vec{v}' \bullet \vec{\varepsilon}'}{c}\right)^2} \qquad (21)$$

and from (21, 6)

$$-\frac{\partial \vec{A}}{\partial T} \wedge \nabla_r T = \frac{q\vec{a}' \wedge \vec{\varepsilon}' + q\vec{\varepsilon}' \wedge \left[\vec{\varepsilon}' \wedge \left(\vec{a}' \wedge \frac{\vec{v}'}{c}\right)\right]}{c^2|\vec{r}-\vec{r}'|\left(1-\frac{\vec{v}' \bullet \vec{\varepsilon}'}{c}\right)^3} + \frac{q\left(\frac{\vec{v}'}{c} \wedge \vec{\varepsilon}'\right)\left[\frac{\vec{v}'}{c} \bullet \left(\vec{\varepsilon}' - \frac{\vec{v}'}{c}\right)\right]}{|\vec{r}-\vec{r}'|^2 \left(1-\frac{\vec{v}' \bullet \vec{\varepsilon}'}{c}\right)^3} \qquad (22)$$

From (17, 20, 22) the magnetic field is given by

$$\vec{H}(\vec{r},t) = \nabla_r \wedge \vec{A} = \frac{q\left(\frac{\vec{v}'}{c} \wedge \vec{\varepsilon}'\right)\left(1-\frac{\vec{v}'^2}{c^2}\right)}{|\vec{r}-\vec{r}'|^2 \left(1-\frac{\vec{v}' \bullet \vec{\varepsilon}'}{c}\right)^3} + \vec{\varepsilon}' \wedge \frac{-q\vec{a}' + q\left[\vec{\varepsilon}' \wedge \left(\vec{a}' \wedge \frac{\vec{v}'}{c}\right)\right]}{c^2|\vec{r}-\vec{r}'|\left(1-\frac{\vec{v}' \bullet \vec{\varepsilon}'}{c}\right)^3} \qquad (23)$$

Equations (15, 23) constitute the electric and magnetic fields for the Lienard-Wiechert potentials. Each of these fields separate out naturally into (a) fields that vary as $1/|\vec{r}-\vec{r}'|^2$ and is independent of the acceleration (b) fields that vary as $1/|\vec{r}-\vec{r}'|$ and involves the acceleration. The latter constitute the radiation fields for a moving charge. They are orthogonal to one another and to $\vec{\varepsilon}'$.

From (15, 23) the radiation field is specified by the electric and magnetic fields

$$\vec{E}^*(\vec{r},t) = \frac{-q[\vec{a}' - (\vec{a}' \bullet \vec{\varepsilon}')\vec{\varepsilon}'] + q\vec{\varepsilon}' \wedge \left(\vec{a}' \wedge \frac{\vec{v}'}{c}\right)}{c^2|\vec{r}-\vec{r}'|\left(1-\frac{\vec{v}' \bullet \vec{\varepsilon}'}{c}\right)^3} \qquad (24)$$

$$\vec{H}^*(\vec{r},t) = \vec{\varepsilon}' \wedge \frac{-q\vec{a}' + q\left[\vec{\varepsilon}' \wedge \left(\vec{a}' \wedge \frac{\vec{v}'}{c}\right)\right]}{c^2 |\vec{r}-\vec{r}'|\left(1-\frac{\vec{v}' \bullet \vec{\varepsilon}'}{c}\right)^3} \qquad (25)$$

Therefore,

$$\vec{E}^* \bullet \vec{\varepsilon}' = \vec{H}^* \bullet \vec{\varepsilon}' = 0 \qquad (26)$$

and a simple manipulation shows,

$$\vec{E}^* \bullet \vec{H}^* = 0 \qquad (27)$$

Since, $\vec{\varepsilon}' \wedge \vec{\varepsilon}' = \vec{0}$, we can re-write (25) as

$$\vec{H}^*(\vec{r},t) = \vec{\varepsilon}' \wedge \frac{-q[\vec{a}' - (\vec{a}' \bullet \vec{\varepsilon}')\vec{\varepsilon}'] + q\left[\vec{\varepsilon}' \wedge \left(\vec{a}' \wedge \frac{\vec{v}'}{c}\right)\right]}{c^2 |\vec{r}-\vec{r}'|\left(1-\frac{\vec{v}' \bullet \vec{\varepsilon}'}{c}\right)^3}$$

$$= \vec{\varepsilon}' \wedge \vec{E}^* \qquad (27\text{-}1)$$

$$\therefore |\vec{H}^*|^2 = (\vec{\varepsilon}' \wedge \vec{E}^*) \bullet (\vec{\varepsilon}' \wedge \vec{E}^*) = \vec{\varepsilon}' \bullet [\vec{E}^* \wedge (\vec{\varepsilon}' \wedge \vec{E}^*)]$$

$$= \vec{\varepsilon}' \bullet [(\vec{E}^* \bullet \vec{E}^*)\vec{\varepsilon}' - (\vec{E}^* \bullet \vec{\varepsilon}')\vec{E}^*] = |\vec{E}^*|^2 \qquad (28) \quad \{\text{using } (26)\}$$

References:

P.G. Bergmann - Introduction to the Theory of Relativity; Prentice-Hall,1950; Chapter 9.
M. Schwartz - Principles of Electrodynamics, McGraw-Hill, 1972; Chapters 5, 6.
F. Mandl - Statistical Physics, John Wiley, 1971, Chapters 6, 10
N. Tralli - Classical Electromagnetic Theory, McGraw-Hill, 1963, Chapter 17

Q and A Section for Ch 2

1. Show that the function $y=y(x)$ that extremizes (i.e. maximizes or minimizes) the functional

$$I[y(x)] = \int_a^b f(x, y(x), y'(x)) dx \;\; ;$$

   with $\qquad y'(x) = \dfrac{dy}{dx}; \;\; y(a) = A; y(b) = B$

   is given by,

$$\frac{\partial f}{\partial x} - \frac{d}{dx}\left(\frac{\partial f}{\partial y'}\right) = 0$$

   where $(a, A), (b, B)$ are constants

2. Extremize the functional,

$$I[y(x_1, x_2, ..., x_n)] = \int f(x_1, x_2, ..., x_n; y(x_1, x_2, ..., x_n); \frac{\partial y}{\partial x_1}, \frac{\partial y}{\partial x_2}, ..., \frac{\partial y}{\partial x_n}) dx_1 dx_2 ... dx_n$$

3. Extremize the functional,

$$I[y_1(x), y_2(x), ..., y_n(x)] = \int f(x; y_1(x), y_2(x), ..., y_n(x); y_1'(x), y_2'(x), ..., y_n'(x)) dx$$

4. Prove that for a curved space time with metric $ds^2 = g_{\mu\nu} dx^\mu dx^\nu$, the geodesics are given by

$$\frac{d^2 x^\rho}{ds^2} + \Gamma^\rho_{\lambda\sigma} \frac{dx^\lambda}{ds} \frac{dx^\sigma}{ds} = 0$$

   where the Γ are the Christoffel symbols of the second kind defined by,

$$\Gamma^\lambda_{\rho\sigma} = \frac{1}{2} g^{\lambda\omega} (g_{\rho\omega,\sigma} + g_{\sigma\omega,\rho} - g_{\rho\sigma,\omega})$$

   using the sign convention, implying summation over a repeated index. A geodesic is the shortest distance between 2 events in 4 dimensional space-time.

5. Determine the geodesics for a flat space-time metric (1).

6. Determine the geodesics for the Schwarzschild metric (7).

7. Hubble's Function for the following universes are defined as

$$H(t) = H_0 \quad \text{constant (Einstein)}$$

$$= \frac{2}{3t} \quad \text{(de Sitter)}$$

$$= at \quad \text{(Milne ?)}$$

Derive Hubble's velocity-distance law for the above universes, equation (24)

8. Derive the mass conversion factor k(t) for the universes in Exercise 7, using the equations (51 - 56)

9. Calculate the trajectory of a classical particle of mass m_0 and charge q moving in a uniform electric and magnetic field of intensities **E, H** respectively. For purposes of simplicity, assume the fields are orthogonal. Show that this trajectory reduces to a spiral helix in the absence of an electric field.

10. A horizontal beam of particles each of mass m_0 and charge q is moving with a uniform velocity u when a uniform magnetic field of strength H is applied normal to the direction of motion. Show that the vertical deflection d is given by,

$$d = \frac{qH}{2m_0 cu} l^2$$

where l is the horizontal distance traveled. Discuss any experimental ramifications of this result.

SOLUTIONS

1.
$$I[y(x)] = \int_a^b f(x, y(x), y'(x))dx$$

$$\therefore \quad I[y + \delta y] = \int_a^b f(x, y + \delta y, y' + \delta y')dx$$

$$\delta I \equiv I[y+\delta y] - I[y] = \int_a^b [f(x, y+\delta y, y'+\delta y') - f(x, y, y')]dx$$

From Taylor's Theorem,

$$\delta I = \int_a^b \left(\frac{\partial f}{\partial y}\delta y - \frac{\partial f}{\partial y'}\delta y'\right)dx$$

Integrating by parts,

$$\int_a^b \frac{\partial f}{\partial y'}\delta y' dx = \left[\frac{\partial f}{\partial y'}\delta y\right] - \int_a^b \delta y \frac{d}{dx}\left(\frac{\partial f}{\partial y'}\right)dx$$

But $\left[\dfrac{\partial f}{\partial y'}\delta y\right] = 0$, because $\delta y = 0 \; @ \; x = a, b$

Therefore,

$$\delta I = \int_a^b \left[\frac{\partial f}{\partial y} - \frac{d}{dx}\left(\frac{\partial f}{\partial y'}\right)\right]\delta y \, dx$$

For an extremum,

$\delta I = 0$ for all δy

$$\therefore \frac{\partial f}{\partial y} - \frac{d}{dx}\left(\frac{\partial f}{\partial y'}\right) = 0 \;\; \rightarrow \text{Euler-Lagrange equations}$$

2. $I[y(x_1, x_2, ..., x_n)] = \int f(x_1, x_2, ..., x_n; y(x_1, x_2, ..., x_n); \dfrac{\partial y}{\partial x_1}, \dfrac{\partial y}{\partial x_2}, ..., \dfrac{\partial y}{\partial x_n}) dx_1 dx_2 ... dx_n$

As before,

$$\delta I = I[y+\delta y] - I[y] = \int [f(\boldsymbol{x}; y(\boldsymbol{x}) + \delta y(\boldsymbol{x}); y_{x_1} + \delta y_{x_1}; y_{x_2} + \delta y_{x_2}, ..., y_{x_n} + \delta y_{x_n})$$
$$- f(\boldsymbol{x}; y(\boldsymbol{x}); y_{x_1}, ..., y_{x_n})] \, dx_1 dx_2 .....dx_n$$

where,

$$x = (x_1, x_2, \ldots, x_n) \quad ; \quad y_{x_r} = \frac{\partial y}{\partial x_r} \quad ; \quad r = 1 \to n$$

$$\delta I = \int [\delta y \frac{\partial f}{\partial y} + \sum_{r=1}^{r=n} \delta y_{x_r} \frac{\partial f}{\partial y_{x_r}}] d\mathbf{x}$$

integrating by parts the second term we have,

$$\delta I = \int \delta y \left(\frac{\partial f}{\partial y} - \sum_{r=1}^{r=n} \frac{\partial}{\partial x_r} (\frac{\partial f}{\partial y_{x_r}}) \right) d\mathbf{x}$$

Therefore,

$\delta I = 0$ for arbitrary δy implies,

$$\frac{\partial f}{\partial y} - \sum_{r=1}^{n} \frac{\partial}{\partial x_r} \left(\frac{\partial f}{\partial y_{x_r}} \right) = 0$$

3. In direct analogy with the examples(1, 2)

$$\frac{\partial f}{\partial y_r} - \frac{d}{dx} \left(\frac{\partial f}{\partial y'_r} \right) = 0 \quad ; \quad r = 1 \to n$$

4. $I[x^1, x^2, x^3, x^4] = \int \sqrt{g_{\mu\nu} dx^\mu dx^\nu} = \int \sqrt{g_{\mu\nu} \dot{x}^\mu \dot{x}^\nu} \, ds = \int \Im(s, x^\mu, \dot{x}^\nu) ds$

where a dot denotes differentiation with respect to distance s in curved space-time and,

$\Im(s, \dot{x}^\mu, \dot{x}^\nu) = \sqrt{g_{\mu\nu} \dot{x}^\mu \dot{x}^\nu}$, using the summation convention. Therefore the Euler-Lagrange equations are,

$$\frac{\partial \Im}{\partial x^\mu} - \frac{d}{ds} \left(\frac{\partial \Im}{\partial \dot{x}^\mu} \right) = 0 \quad ; \quad \mu = 1,2,3,4$$

Therefore,

$$\frac{\partial \Im}{\partial x^\mu} = \frac{\dot{x}^\alpha \dot{x}^\beta}{2\Im} g_{\alpha\beta,\mu} = \frac{\dot{x}^\alpha \dot{x}^\beta}{2} g_{\alpha\beta,\mu}$$

$$\frac{\partial \Im}{\partial \dot{x}^\mu} = \frac{g_{\mu\alpha}\dot{x}^\alpha}{\Im} = g_{\mu\alpha}\dot{x}^\alpha$$

because along a geodesic,

$$ds = \sqrt{g_{\mu\nu}dx^\mu dx^\nu} = \sqrt{g_{\mu\nu}\frac{dx^\mu}{ds}\frac{dx^\nu}{ds}}ds = \sqrt{g_{\mu\nu}\dot{x}^\mu \dot{x}^\nu}\,ds = \Im\, ds$$

∴
$$\Im = 1$$

The chain rule of partial differentiation gives along a geodesic,

$$\frac{d\Im}{ds} = \dot{x}^\alpha \frac{\partial \Im}{\partial x^\alpha} + \ddot{x}^\beta \frac{\partial \Im}{\partial \dot{x}^\beta} = \dot{x}^\rho \dot{x}^\lambda \dot{x}^\sigma g_{\lambda\sigma,\rho}/2 + \ddot{x}^\beta g_{\beta\tau}\dot{x}^\tau$$

$$\frac{dg_{\mu\nu}}{ds} = \frac{dx^\beta}{ds}\frac{dg_{\mu\nu}}{dx^\beta} = \dot{x}^\beta g_{\mu\nu,\beta}$$

∴ $$\frac{d}{ds}\left(\frac{\partial \Im}{\partial \dot{x}^\mu}\right) = \frac{d}{ds}\left(g_{\mu\alpha}\dot{x}^\alpha\right) = g_{\mu\alpha}\ddot{x}^\alpha + \dot{x}^\alpha \frac{dg_{\mu\alpha}}{ds}$$

$$= g_{\mu\alpha}\ddot{x}^\alpha + \dot{x}^\alpha \dot{x}^\beta g_{\mu\alpha,\beta}$$

∴ $$\frac{\partial \Im}{\partial x^\mu} - \frac{d}{ds}\left(\frac{\partial \Im}{\partial \dot{x}^\mu}\right) = \frac{\dot{x}^\alpha \dot{x}^\beta}{2}g_{\alpha\beta,\mu} - g_{\mu\alpha}\ddot{x}^\alpha - \dot{x}^\alpha \dot{x}^\beta g_{\mu\alpha,\beta}$$

$$= \left(\frac{1}{2}g_{\alpha\beta,\mu} - g_{\mu\alpha,\beta}\right)\dot{x}^\alpha \dot{x}^\beta - g_{\mu\alpha}\ddot{x}^\alpha$$

$$= \left(\frac{1}{2}g_{\alpha\beta,\mu} - g_{\mu\beta,\alpha}\right)\dot{x}^\alpha \dot{x}^\beta - g_{\mu\alpha}\ddot{x}^\alpha$$

because the dummy suffices α, β inside the brackets are interchangeable in accordance with the summation convention and $g_{\alpha\beta} = g_{\beta\alpha}$, the metric tensor $g_{\mu\nu}$ is symmetric.

∴ $$\frac{\partial \Im}{\partial x^\mu} - \frac{d}{ds}\left(\frac{\partial \Im}{\partial \dot{x}^\mu}\right) = \frac{1}{2}\left(g_{\alpha\beta,\mu} - g_{\mu\alpha,\beta} - g_{\mu\beta,\alpha}\right)\dot{x}^\alpha \dot{x}^\beta - g_{\mu\alpha}\ddot{x}^\alpha$$

Therefore, the Euler-Lagrange equations become,

$$\frac{1}{2}(g_{\alpha\beta,\mu} - g_{\mu\alpha,\beta} - g_{\mu\beta,\alpha})\dot{x}^\alpha \dot{x}^\beta - g_{\mu\alpha}\ddot{x}^\alpha = 0$$

Multiplying both sides of this equation by $g^{\rho\mu}$ and summing with respect to the repeated suffix μ,

$$\frac{g^{\rho\mu}}{2}(g_{\alpha\beta,\mu} - g_{\mu\alpha,\beta} - g_{\mu\beta,\alpha})\dot{x}^\alpha \dot{x}^\beta - g^{\rho\mu}g_{\mu\alpha}\ddot{x}^\alpha = 0$$

i.e.

$$\frac{g^{\rho\mu}}{2}(g_{\alpha\beta,\mu} - g_{\mu\alpha,\beta} - g_{\mu\beta,\alpha})\dot{x}^\alpha \dot{x}^\beta - \delta^\rho_\alpha \ddot{x}^\alpha = 0$$

where δ^ρ_α is the Kronecker delta. This reduces to,

$$\frac{g^{\rho\mu}}{2}(g_{\alpha\beta,\mu} - g_{\mu\alpha,\beta} - g_{\mu\beta,\alpha})\dot{x}^\alpha \dot{x}^\beta - \ddot{x}^\rho = 0$$

$$\therefore \quad \ddot{x}^\rho + \frac{g^{\rho\mu}}{2}(g_{\mu\alpha,\beta} + g_{\mu\beta,\alpha} - g_{\alpha\beta,\mu})\dot{x}^\alpha \dot{x}^\beta = 0$$

$$\therefore \quad \ddot{x}^\rho + \frac{g^{\rho\mu}}{2}(g_{\mu\lambda,\sigma} + g_{\mu\sigma,\lambda} - g_{\lambda\sigma,\mu})\dot{x}^\lambda \dot{x}^\sigma = 0$$

$$\therefore \quad \ddot{x}^\rho + \frac{g^{\rho\omega}}{2}(g_{\omega\lambda,\sigma} + g_{\omega\sigma,\lambda} - g_{\lambda\sigma,\omega})\dot{x}^\lambda \dot{x}^\sigma = 0$$

$$\therefore \quad \frac{d^2 x^\rho}{ds^2} + \Gamma^\rho_{\lambda\sigma}\frac{dx^\lambda}{ds}\frac{dx^\sigma}{ds} = 0, \text{ because } \mu,\omega \text{ are dummy suffices.}$$

5. Since all the metric coefficients are either ± 1, their derivatives are zero. Therefore, $\Gamma^\rho_{\lambda\sigma} = 0$, for all $\rho, \lambda, \sigma = 1, 2, 3, 4$.

$$\therefore \quad \frac{d^2 x^\rho}{ds^2} = 0 \rightarrow \text{relativistic analog of Newton's First and (Second) Law of motion.}$$

Integrating,

$$\frac{x^1 - a^1}{l^1} = \frac{x^2 - a^2}{l^2} = \frac{x^3 - a^3}{l^3} = \frac{x^4 - a^4}{l^4} = s$$

where $(a^1, a^2, a^3, a^4), (l^1, l^2, l^3, l^4)$ are constants of integration. This represents a straight line (world line) in 4 dimensions.

Force : $F^\rho \approx c\dfrac{dp^\mu}{ds}$ (SP)

Special Relativity:

$$x^\mu = (x,y,z,t) = (\mathbf{r},t) \; ; \; u^\mu = c\dfrac{dx^\mu}{ds} \; ; \; p^\mu = m_0 u^\mu$$

$$\eta_{\mu\nu} = 0, \mu \neq \nu \; ; \; \eta_{11} = \eta_{22} = \eta_{33} = -1 \; ; \; \eta_{44} = c^2$$

$$\eta_{\mu\nu} u^\mu u^\nu = c^2$$

General relativity : Replace $\eta_{\mu\nu} \to g_{\mu\nu}$

6 . Schwarzschild metric : $ds^2 = c^2 \gamma dt^2 - \dfrac{dr^2}{\gamma} - r^2 d\theta^2 - r^2 \sin^2\theta d\phi^2$

$$\gamma = 1 - \dfrac{2MG}{rc^2}$$

$\therefore g_{\mu\nu} = 0, \mu \neq \nu \; ;$

$$g_{rr} = -\dfrac{1}{\gamma} ; g_{\theta\theta} = -r^2 ; g_{\phi\phi} = -r^2 \sin^2\theta ; g_{tt} = c^2\gamma$$

$g^{\mu\nu} = 0 ; \mu \neq \nu \; ;$

$$g^{rr} = -\gamma ; g^{\theta\theta} = -\dfrac{1}{r^2} ; g^{\phi\phi} = -\dfrac{1}{r^2 \sin^2\theta} ; g^{tt} = \dfrac{1}{c^2\gamma}$$

Then only nonzero partial derivatives of the metric tensor are:

$$g_{rr,r} = \dfrac{2MG}{r^2 c^2 \gamma^2}$$

$$g_{\theta\theta,r} = -2r$$

$$g_{\phi\phi,r} = -2r\sin^2\theta \; ; \; g_{\phi\phi,\theta} = -2r^2 \sin\theta \cos\theta$$

$$g_{tt,r} = \dfrac{2MG}{r^2}$$

The only nonzero Christoffel symbols ($\Gamma^\lambda_{\mu\nu} = \Gamma^\lambda_{\nu\mu}$) are (no summation)

$$\Gamma^r_{rr} = \frac{1}{2} g^{rr} g_{rr,r} = -\frac{MG}{r^2 c^2 \gamma}$$

$$\Gamma^r_{\theta\theta} = -\frac{1}{2} g^{rr} g_{\theta\theta,r} = -r\gamma$$

$$\Gamma^r_{\phi\phi} = -\frac{1}{2} g^{rr} g_{\phi\phi,r} = -\gamma r \sin^2 \theta$$

$$\Gamma^r_{tt} = -\frac{1}{2} g^{rr} g_{tt,r} = \frac{MG}{r^2} \gamma$$

$$\Gamma^\theta_{\theta r} = \frac{1}{2} g^{\theta\theta} g_{\theta\theta,r} = \frac{1}{r}$$

$$\Gamma^\theta_{\phi\phi} = -\frac{1}{2} g^{\theta\theta} g_{\phi\phi,\theta} = \sin\theta \cos\theta$$

$$\Gamma^\phi_{r\phi} = \frac{1}{2} g^{\phi\phi} g_{\phi\phi,r} = \frac{1}{r}$$

$$\Gamma^t_{rt} = \frac{1}{2} g^{tt} g_{tt,r} = \frac{MG}{\gamma r^2 c^2}$$

Therefore, the geodesics for for Schwarzschild metric are given by,

$$\frac{d^2 r}{ds^2} + \Gamma^r_{rr} \left(\frac{dr}{ds}\right)^2 + \Gamma^r_{\theta\theta} \left(\frac{d\theta}{ds}\right)^2 + \Gamma^r_{\phi\phi} \left(\frac{d\phi}{ds}\right)^2 + \Gamma^r_{tt} \left(\frac{dt}{ds}\right)^2 = 0$$

$$\frac{d^2 \theta}{ds^2} + 2\Gamma^\theta_{\theta r} \left(\frac{d\theta}{ds}\right)\left(\frac{dr}{ds}\right) + \Gamma^\theta_{\phi\phi} \left(\frac{d\phi}{ds}\right)^2 = 0$$

$$\frac{d^2 \phi}{ds^2} + 2\Gamma^\phi_{r\phi} \left(\frac{dr}{ds}\right)\left(\frac{d\phi}{ds}\right) = 0$$

$$\frac{d^2 t}{ds^2} + 2\Gamma^t_{rt} \left(\frac{dr}{ds}\right)\left(\frac{dt}{ds}\right) = 0$$

Substituting for the Γ's

$$\frac{d^2r}{ds^2} - \frac{MG}{r^2c^2\gamma}\left(\frac{dr}{ds}\right)^2 - r\gamma\left(\frac{d\theta}{ds}\right)^2 - r\gamma\sin^2\theta\left(\frac{d\phi}{ds}\right)^2 - \frac{MG\gamma}{r^2}\left(\frac{dt}{ds}\right)^2 = 0$$

$$\frac{d^2\theta}{ds^2} + \frac{2}{r}\left(\frac{d\theta}{ds}\right)\left(\frac{dr}{ds}\right) + \sin\theta\cos\theta\left(\frac{d\phi}{ds}\right)^2 = 0$$

$$\frac{d^2\phi}{ds^2} + \frac{2}{r}\left(\frac{dr}{ds}\right)\left(\frac{d\phi}{ds}\right) = 0$$

$$\frac{d^2t}{ds^2} + 2\frac{MG}{\gamma r^2 c^2}\left(\frac{dr}{ds}\right)\left(\frac{dt}{ds}\right) = 0$$

7. From equations (23, 24),   Hubble Parameter → Einstein Universe

Einstein Universe: $H(t) = H_0$

$r(t) = r_0 e^{H_0(t-t_0)}$

$v(t) = H_0 r(t)$

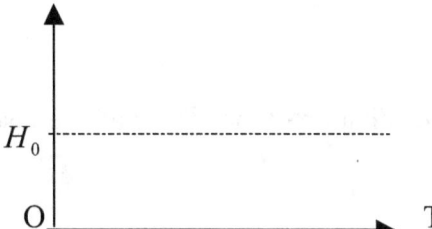

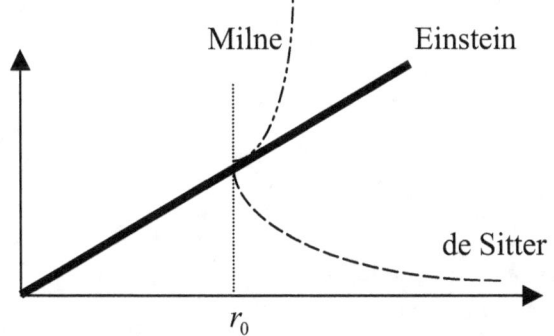

Present Time t_0

de Sitter Universe: $H(t) = \dfrac{2}{3t}$

Hubble Parameter → de Sitter Universe

$r(t) = r_0 \left(\dfrac{t}{t_0} \right)^{2/3}$

$v(t) = H(t) r(t)$

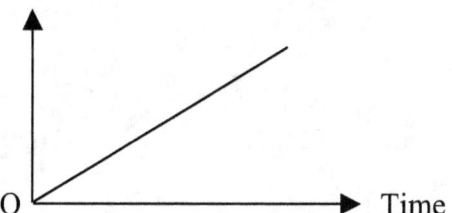

Hubble Parameter → Milne Universe

Milne Universe: $H(t) = at\,?$

$r(t) = r_0 e^{\frac{a}{2}(t^2 - t_0^2)}$

$v(t) = H(t) r(t)$

8. A direct substitution of the results of exercise 7 in equation 55, gives the radiation to mass ratio $k(t)$.

Einstein Universe: $r(t) = r_0 e^{H_0 (t - t_0)}$

$v(t) = H_0 r_0 e^{H_0 (t - t_0)}$

Since, $v(t) < c \quad \Rightarrow \quad e^{H_0(t-t_0)} < \dfrac{c}{H_0 r_0}$

de Sittrer Universe: $H(t) = \dfrac{2}{3t}$

$r(t) = r_0 \left(\dfrac{t}{t_0} \right)^{2/3}$

$v(t) = \dfrac{2}{3 t^{\frac{1}{3}} t_0^{\frac{2}{3}}} r_0$

Since, $$\frac{2}{3t^{\frac{1}{3}}t_0^{\frac{2}{3}}}r_0 < c$$

These requirements impose restrictions on the time t.

9. Without loss of generality, let us assume that the electric field is along the Ox axis while the magnetic field is parallel to the Oz axis.

$\mathbf{E} = E_0\, \mathbf{i}$; $\mathbf{H} = H_0\, \mathbf{k}$; $\mathbf{i}, \mathbf{j}, \mathbf{k}$ = unit vectors along Ox, Oy, Oz respectively.

The equation of motion is given by,

$$m_0 \frac{d}{dt}\mathbf{v} = q(\mathbf{E} + \frac{1}{c}\mathbf{v} \wedge \mathbf{H}) \qquad \text{Lorentz force}$$

in component form,

$$m_0 \frac{d^2 x}{dt^2} = q(E_0 + \frac{H_0}{c}\frac{dy}{dt})$$

$$m_0 \frac{d^2 y}{dt^2} = -q\frac{H_0}{c}\frac{dx}{dt}$$

$$m_0 \frac{d^2 z}{dt^2} = 0$$

Put, $\dot{\xi} = \frac{dx}{dt} + i\frac{dy}{dt}$,

Therefore,

$$m_0 \frac{d\dot{\xi}}{dt} = qE_0 - i\frac{qH_0}{c}\dot{\xi}$$

$\therefore \quad \frac{d\dot{\xi}}{dt} + i\frac{qH_0}{m_0 c}\dot{\xi} = \frac{qE_0}{m_0}$

Integrating,

$$\dot{\xi} = \dot{\xi}_0 e^{-i\frac{qH_0}{m_0 c}t} - \frac{icE_0}{H_0}\left(1 - e^{-i\frac{qH_0}{m_0 c}t}\right)$$

where at $t = 0, \dot{\xi} = \dot{\xi}_0$. Integrating again,

$$\xi = \xi_0 + \frac{m_0 c}{i q H_0}\dot{\xi}_0(1-e^{-i\frac{qH_0}{m_0 c}t}) + \frac{m_0 c^2}{H_0^2}E_0(1-e^{-i\frac{qH_0}{m_0 c}t}) - i\frac{cE_0}{H_0}t$$

where at $t = 0, \xi = \xi_0$. Equating real and imaginary parts,

$$x = x_0 + \frac{m_0 c}{qH_0}\left[v_0(1-\cos\omega t) + u_0 \sin\omega t\right] + \frac{m_0 c^2}{H_0}E_0(1-\cos\omega t)$$

$$y = y_0 - \frac{m_0 c}{qH_0}\left[u_0(1-\cos\omega t) - v_0 \sin\omega t\right] + \frac{m_0 c^2}{H_0}E_0 \sin\omega t - \frac{cE_0}{H_0}t$$

where,

$$\xi_0 = x_0 + iy_0 \;;\; \dot{\xi} = u_0 + iv_0 \;\&\; \omega = \frac{qH_0}{m_0 c}$$

In the absence of an electric field the above equations reduce to,

$$x = x_0 + \frac{m_0 c}{qH_0}\left[v_0(1-\cos\omega t) + u_0 \sin\omega t\right]$$

$$y = y_0 - \frac{m_0 c}{qH_0}\left[u_0(1-\cos\omega t) - v_0 \sin\omega t\right]$$

Therefore,

$$(x-x_0)^2 + (y-y_0)^2 = 4\left(\frac{m_0 c}{qH_0}\right)^2 (u_0^2 + v_0^2)\sin^2\left(\frac{\omega t}{2}\right)$$

$$\leq 4\left(\frac{m_0 c}{qH_0}\right)^2 (u_0^2 + v_0^2)$$

$$\frac{dz}{dt} = w_0 \Rightarrow z = z_0 + w_0 t$$

10. Kinematics : uniform acceleration f

$$v = u + ft$$

$$s = ut + \frac{1}{2}ft^2$$

$$v^2 = u^2 + 2fs$$

where, u is the initial velocity of a particle and v is its velocity at the end of time t s is the distance traveled.

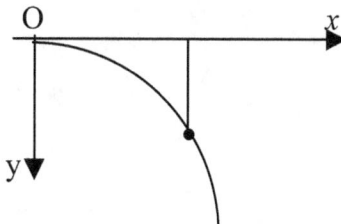

The particle is traveling horizontally with a constant velocity u, when a uniform magnetic field H is applied normal to the plane of the paper, pointing towards the reader. The particle will experience a downward force in the Oy direction as shown.

Horizontal equation of motion $\qquad x = ut$

Vertical force acting on the particle $\qquad = \dfrac{quH}{c}$

Vertical acceleration $\qquad f = \dfrac{quH}{m_0 c}$

Vertical equation of motion $\qquad y = \dfrac{1}{2} ft^2 \;\rightarrow$ no initial vertical velocity

Therefore eliminating t

$$y = \frac{1}{2} f \left(\frac{x}{u}\right)^2 = \frac{qH}{2m_0 cu} x^2$$

Experimental ramifications : Aston's mass spectrograph used in nuclear physics. If an electric field is now applied in the plane of the paper, so as to pull the beam up, till the beam is once again horizontal. Then the electric and magnetic forces are in balance.

Q & A Section for Chapter 3

1. Use the matrix methods of Heisenberg to determine the energy levels for a harmonic oscillator.

2. A pure radiation field is enclosed in a cube of side L, and volume V with periodic boundary conditions. Show that the electromagnetic field can be represented by a series of harmonic oscillators.

3. Discuss mathematically the trembling motion or zitterbewegung associated with the measurement of the position of a particle in the Heisenberg picture. Use the Dirac equation for the one dimensional motion of an electron to highlight the phenomenon.

4. Determine the solutions to Dirac's wave equation in the absence of any external force fields.

5. Write down the matrix elements for the emission and absorption of photons by an electron using Dirac's theory.

6. Use the partition function for a canonical ensemble to determine the (a) the equation of state (b) Maxwell- Boltzmann velocity distribution for a perfect gas.

7. Derive the Virial Theorem of statistical mechanics. Use it to derive Van der Waals equation of state for a weakly interacting monatomic gas.

8. The problem of the calculation of the most probable distribution for a dynamical system may be posed as follows:

   Maximize the probability function,

   $$P = \frac{N!}{a_1! a_2! a_3! ... a_l! ..}$$

   subject to the conditions: $\sum_l a_l = N$; $\sum_l \varepsilon_l a_l = E$.

   There are N identical systems of which $a_1, a_2, a_3, ..., a_l, ...$ are in states labeled $(1, 2, 3, ..., l, ...)$, with energies $(\varepsilon_1, \varepsilon_2, \varepsilon_3, ..., \varepsilon_l, ...)$ respectively.

   Show that : $a_l = N \dfrac{e^{-\beta \varepsilon_l}}{\sum_l e^{-\beta \varepsilon_l}}$, $\beta = \dfrac{1}{kT}$, k = Boltzmann's constant

$$\frac{E}{N} = \frac{\sum_{l} \varepsilon_{l} e^{-\beta \varepsilon_{l}}}{\sum_{l} e^{-\beta \varepsilon_{l}}}$$

9. Derive Stirling's formula for large n

   $$\log n! = n(\log n - 1)$$

   Derive Nernst's formula for the limit of the entropy as the absolute temperature tends to zero is a universal constant.

   $$\underset{T \to 0}{Lt} \ S = k \log g_1, \text{ where } g_1 = \text{ degeneracy of the lowest state.}$$

   Discuss the significance of $g_1 = 1$.

10. (a) Show that 1 gm atom of an element of the periodic table always contains the same number of atoms (Avogadro's number).

    (b) Prove by induction or otherwise, the multinomial theorem for a positive integral index N.

    $$(x_1 + x_2 + x_3 + ... + x_l + ...)^N = \sum_{a_l} \frac{N!}{a_1! a_2! a_3! ... a_l! ...} x_1^{a_1} x_2^{a_2} x_3^{a_3} .... x_l^{a_l} ...$$

    the sum being taken over all integer subsets $(a_1, a_2, a_3, ..., a_l, ...)$ consistent with,

    $$\sum_{l} a_l = N$$

Solutions:

1. $\dfrac{df}{dt} = \dfrac{i}{\hbar}[H, f] = \dfrac{i}{\hbar}[Hf - fH]$

   Harmonic oscillator of frequency v : $\quad \ddot{q} + 4\pi^2 v^2 q^2 = 0$ \hfill (1)

   Total Energy : $\quad H = \dfrac{p^2}{2m} + 2\pi^2 v^2 m q^2 \ ; \quad p = m\dot{q}$ \hfill (2)

   $$\dot{q} = \frac{i}{\hbar}[Hq - qH] = \frac{i}{2m\hbar}[p^2 q - qp^2] = \frac{i}{2m\hbar}[p(pq - qp) + (pq - qp)p]$$

$$= \frac{i}{2m\hbar}\{-p[q,p]-[q,p]p\} = \frac{i}{2m\hbar}[-i\hbar p - i\hbar p] = \frac{p}{m} \qquad (3)$$

$$\dot{p} = \frac{i}{\hbar}[Hp - pH] = \frac{2\pi^2 v^2 mi}{\hbar}[q^2 p - pq^2] = \frac{2\pi^2 v^2 mi}{\hbar}[q(qp-pq)+(qp-pq)q]$$

$$= \frac{2\pi^2 v^2 mi}{\hbar}\{q[q,p]+[q,p]q\} = \frac{2\pi^2 v^2 mi}{\hbar}[i\hbar q + i\hbar q] = -4\pi^2 v^2 mq \qquad (4)$$

$p, q, H,..$ are matrices in an infinite dimensional vector space, with elements $\{p_{lk}, q_{lk}, H_{lk}\}$ respectively. In addition, these matrices are self-adjoint. From (3,4)

$$\ddot{q} + 4\pi^2 v^2 q = 0 \text{ , as expected.} \Rightarrow \ddot{q}_{lk} + 4\pi^2 v^2 q_{lk} = 0 \qquad (5)$$

Try solutions of the form,

$$q_{lk} = a_{lk} e^{-2\pi i v_{lk} t} \qquad (5a)$$

Substituting (5a) in (5),

$$(v^2 - v_{lk}^2) q_{lk} = 0 \qquad (5b)$$

$$\therefore \qquad q_{lk} = 0 \text{ , if } v_{lk} \neq \pm v \qquad (5c)$$

without loss of generality, we take

$v_{n-1,n} = v$ → emission of a quantum of energy hv

$v_{n,n-1} = -v$ → absorption of a quantum of energy hv

(5d)

So that the only nonzero elements of the q − matrix are : $q_{n-1,n}, q_{n,n-1}$, that is $q_{lk} \neq 0$, if and only if $|l-k|=1$.

From (2, 5a, 5c) the only non-vanishing elements of the momentum matrix are,

$$p_{lk} = -2\pi i v_{lk} m q_{lk} ; \Rightarrow p_{n-1,n} = -2\pi i m v q_{n-1,n} ; p_{n,n-1} = 2\pi i m v q_{n,n-1} \qquad (5e)$$

Therefore,

$$q = \begin{pmatrix} 0, q_{01}, 0, 0, \dots \\ q_{10}, 0, q_{12}, 0, \dots \\ 0, q_{21}, 0, q_{23}, \dots \\ \dots \\ \dots \end{pmatrix} \quad , \quad p = 2\pi i m \nu \begin{pmatrix} 0, -q_{01}, 0, 0, \dots \\ q_{10}, 0, -q_{12}, 0, \dots \\ 0, q_{21}, 0, -q_{23}, \dots \\ \dots \\ \dots \end{pmatrix} \tag{5f}$$

$$(qp - pq)_{lk} = (qp)_{lk} - (pq)_{lk} = \sum_{\alpha} q_{l\alpha} p_{\alpha k} - \sum_{\alpha} p_{l\alpha} q_{\alpha k}$$

$$= (q_{l,l-1} p_{l-1,k}) + (q_{l,l+1} p_{l+1,k}) - [(p_{l,l-1} q_{l-1,k}) + (p_{l,l+1} q_{l+1,k})]$$

$$= (q_{l,l-1} p_{l-1,k}) + (q_{l,l+1} p_{l+1,k}) - [(p_{l,l-1} q_{l-1,k}) + (p_{l,l+1} q_{l+1,k})] \tag{5g}$$

where $\alpha = l \pm 1$, $k = l-2, l$ for non-zero elements. Therefore, the matrix (5g) is diagonal, with diagonal elements given by (no summation),

$$(qp - pq)_{l,l-2} = (q_{l,l-1} p_{l-1,l-2}) + (q_{l,l+1} p_{l+1,l-2}) - [(p_{l,l-1} q_{l-1,l-2}) + (p_{l,l+1} q_{l+1,l-2})]$$

$$= (q_{l,l-1} p_{l-1,l-2}) - (p_{l,l-1} q_{l-1,l-2}) = 2\pi i m \nu [(q_{l,l-1} q_{l-1,l-2}) - (q_{l,l-1} q_{l-1,l-2})] = 0$$

$$(qp - pq)_{ll} = (q_{l,l-1} p_{l-1,l}) + (q_{l,l+1} p_{l+1,l}) - [(p_{l,l-1} q_{l-1,l}) + (p_{l,l+1} q_{l+1,l})]$$

$$= 2\pi i m \nu [-q_{l,l-1} q_{l-1,l} + q_{l,l+1} q_{l+1,l} - q_{l,l-1} q_{l-1,l} + q_{l,l+1} q_{l+1,l}]$$

$$= 4\pi i m \nu (q_{l,l+1} q_{l+1,l} - q_{l,l-1} q_{l-1,l}) \tag{5h}$$

because from (5c - 5f), the non-zero elements of p are < 0, if the first subscript is less than the second subscript. Therefore,

$$4\pi i m \nu (q_{l,l+1} q_{l+1,l} - q_{l,l-1} q_{l-1,l}) = i\hbar \quad \Rightarrow \quad q_{l,l+1} q_{l+1,l} - q_{l,l-1} q_{l-1,l} = \frac{\hbar}{4\pi m \nu}$$

Adding the sequence (5h), we have

$$q_{n,n+1} q_{n+1,n} = (n+1) \frac{\hbar}{4\pi m \nu} \tag{5i}$$

Since the matrix q is Hermitian, $\quad q_{lk} = q_{kl}^*$ \hfill (5j)

$$|q_{n,n+1}|^2 = (n+1) \frac{\hbar}{4\pi m \nu} \tag{5k}$$

Without loss of generality take,

$$q_{n,n+1} = \sqrt{\frac{(n+1)\hbar}{4\pi m \nu}} e^{-2\pi i \nu t} \quad ; \quad q_{n+1,n} = \sqrt{(n+1)\hbar \over 4\pi m \nu} e^{2\pi i \nu t} \tag{5l}$$

Again the nonzero representation for the q, p matrices are,

$$(q^2)_{lk} = (qq)_{lk} = \sum_\alpha q_{l\alpha} q_{\alpha k} = q_{l,l-1} q_{l-1,k} + q_{l,l+1} q_{l+1,k} \quad ; \quad \alpha = l \pm 1 \tag{5m}$$

$$(q^2)_{l,l-2} = q_{l,l-1} q_{l-1,l-2} + q_{l,l+1} q_{l+1,l-2} = q_{l,l-1} q_{l-1,l-2}$$

$$(q^2)_{l,l} = q_{l,l-1} q_{l-1,l} + q_{l,l+1} q_{l+1,l}$$

$$(p^2)_{lk} = (pp)_{lk} = \sum_\alpha p_{l\alpha} p_{\alpha k} = p_{l,l-1} p_{l-1,k} + p_{l,l+1} p_{l+1,k} \quad ; \quad \alpha = l \pm 1 \tag{5n}$$

$$(p^2)_{l,l-2} = p_{l,l-1} p_{l-1,l-2} + p_{l,l+1} p_{l+1,l-2} = p_{l,l-1} p_{l-1,l-2} = -4\pi^2 \nu^2 m^2 q_{l,l-1} q_{l-1,l-2}$$

$$(p^2)_{l,l} = p_{l,l-1} p_{l-1,l} + p_{l,l+1} p_{l+1,l} = p_{l,l-1} p_{l-1,l-2} = -4\pi^2 \nu^2 m^2 (-q_{l,l-1} q_{l-1,l} - q_{l,l+1} q_{l+1,l})$$

Using the above results give for the matrix representation of the Hamiltonian (2),

$$H_{l,l-2} = \frac{p_{l,l-2}^2}{2m} + 2\pi^2 \nu^2 m q_{l,l-2}^2 = 0 \tag{5o}$$

$$H_{l,l} = \frac{p_{l,l}^2}{2m} + 2\pi^2 \nu^2 m q_{l,l}^2 = 4\pi^2 \nu^2 m (q_{l,l-1} q_{l-1,l} + q_{l,l+1} q_{l+1,l}) \tag{5p}$$

Therefore, H is a diagonal matrix. The energy levels are the eigenvalues of H.

$$E_n = H_{nn} = 4\pi^2 \nu^2 m (q_{n,n-1} q_{n-1,n} + q_{n,n+1} q_{n+1,n}) \tag{6}$$

From (5l),

$$E_n = 4\pi^2 \nu^2 m \frac{\hbar}{4\pi m \nu}(n+n+1) = h\nu(n+\frac{1}{2}) \tag{7}$$

2. With the usual notation Maxwell's Equations can be written in terms of the vector and

scalar potentials ($\vec{A}(\vec{r},t), \phi(\vec{r},t)$), satisfying the relations,

$$\vec{E} = -\vec{\nabla}\phi - \frac{1}{c}\frac{\partial \vec{A}}{\partial t} \quad ; \quad \vec{H} = \vec{\nabla} \wedge \vec{A} \tag{8}$$

together with the Gauge condition,

$$\vec{\nabla} \bullet \vec{A} + \frac{1}{c}\frac{\partial \phi}{\partial t} = 0 \tag{8a}$$

We can write these relations in a slightly more convenient form thru' the transformation equations,

$$\vec{A} = \vec{\mathrm{A}} + \vec{\nabla}\chi \quad , \quad \phi = \Phi - \frac{1}{c}\frac{\partial \chi}{\partial t} \tag{8b}$$

Substituting (8b) in (8) and simplifying,

$$\vec{E} = -\vec{\nabla}\Phi - \frac{1}{c}\frac{\partial \vec{\mathrm{A}}}{\partial t} \quad , \quad \vec{H} = \vec{\nabla} \wedge \vec{\mathrm{A}} \tag{8c}$$

and from (8a),

$$\vec{\nabla} \bullet \vec{\mathrm{A}} = \vec{\nabla} \bullet \vec{A} - \nabla^2 \chi$$

Choose χ such that,

$$\vec{\nabla} \bullet \vec{A} = \nabla^2 \chi \tag{8d}$$

Therefore the modified vector potential satisfies the condition,

$$\vec{\nabla} \bullet \vec{\mathrm{A}} = 0 \tag{8e}$$

But,

$$\vec{\nabla} \bullet \vec{E} = 4\pi\rho \quad , \quad \Rightarrow \nabla^2 \Phi = -4\pi\rho \tag{8f}$$

This is the usual result of electrostatics for a potential Φ and charge density ρ. In the absence of a charge distribution, we can take without loss of generality,

$$\Phi = 0 \tag{8g}$$

Again,

$$\vec{\nabla} \wedge \vec{H} = \frac{4\pi}{c}\vec{j} + \frac{1}{c}\frac{\partial \vec{E}}{\partial t} \quad , \quad \vec{j} = \rho\vec{v} \quad , \quad \vec{v} = \text{velocity of a charged particle} \tag{8h}$$

i.e., $$\vec{\nabla} \wedge (\vec{\nabla} \wedge \vec{A}) = \frac{4\pi}{c}\vec{j} + \frac{1}{c}\frac{\partial}{\partial t}\left(-\vec{\nabla}\Phi - \frac{1}{c}\frac{\partial \vec{A}}{\partial t}\right)$$

$$\therefore \quad \vec{\nabla}(\vec{\nabla} \bullet \vec{A}) - \nabla^2 \vec{A} = \frac{4\pi}{c}\vec{j} - \frac{1}{c}\vec{\nabla}\left(\frac{\partial \Phi}{\partial t}\right) - \frac{1}{c^2}\frac{\partial^2 \vec{A}}{\partial t^2}$$

$$\therefore \quad \nabla^2 \vec{A} - \frac{1}{c^2}\frac{\partial^2 \vec{A}}{\partial t^2} = -\frac{4\pi}{c}\vec{j} + \frac{1}{c}\vec{\nabla}\left(\frac{\partial \Phi}{\partial t}\right) + \vec{\nabla}(\vec{\nabla}\bullet\vec{A})$$

$$= -\frac{4\pi}{c}\vec{j} + \frac{1}{c}\vec{\nabla}\left(\frac{\partial \Phi}{\partial t}\right) \quad , \text{ from (8e)} \tag{8i}$$

To summarize, in the absence of charge or current distribution, $\rho = 0, \vec{j} = \vec{0}$ the modified vector potential satisfies the differential equation,

$$\nabla^2 \vec{A} - \frac{1}{c^2}\frac{\partial^2 \vec{A}}{\partial t^2} = \vec{0} \tag{8j}$$

with the electric and magnetic fields given by,

$$\vec{E} = -\frac{1}{c}\frac{\partial \vec{A}}{\partial t} \quad , \quad \vec{H} = \vec{\nabla} \wedge \vec{A} \tag{8k}$$

Using (8g) in (8i). Using a separation of variable technique it can be easily shown that,

$$\vec{A}(\vec{r},t) = \frac{1}{\sqrt{V}}\sum_l \vec{a}_l[q_l(t)e^{i(\vec{k}_l \bullet \vec{r})} + q_l^*(t)e^{-i(\vec{k}_l \bullet \vec{r})}] \tag{9}$$

with,

$$q_l(t) = e^{-2\pi i v_l t} \tag{9a}$$

$$\vec{k}_l = \frac{2\pi}{L}(n_x, n_y, n_z) \tag{9b}$$

where (n_x, n_y, n_z) are integers, because of the boundary conditions. In addition, $\vec{a}_l$ is a constant vector. From (9),

$$\vec{A}(\vec{r},t) = \frac{1}{\sqrt{V}} \sum_l \vec{a}_l [Q_l(t)\cos(\vec{k}_l \bullet \vec{r}) - \frac{P_l}{2\pi v_l}\sin(\vec{k}_l \bullet \vec{r})] \qquad (9c)$$

where,

$$Q_l = q_l + q_l^* \quad ; \quad P_l = -2\pi i v_l (q_l - q_l^*) \qquad (9d)$$

From (9a, 9d) Q_l, P_l are real and,

$$P_l = \dot{Q}_l \quad ; \quad -4\pi^2 v_l^2 Q_l = \dot{P}_l \qquad (9e)$$

Define,

$$H_l = \frac{1}{2}P_l^2 + 2\pi^2 v_l^2 Q_l^2 \qquad (9f)$$

From (9a, 9e, 9f)

$$\dot{P}_l = -\frac{\partial H_l}{\partial Q_l} \quad ; \quad \dot{Q}_l = \frac{\partial H_l}{\partial P_l} \qquad (9g)$$

Therefore, Q_l, P_l have the status of generalized coordinates and momenta, for the Hamiltonian (9). Comparison with (2) shows (9f) is the Hamiltonian for a harmonic Oscillator with mass = 1. From (9e, 9d, 9a)

$$\ddot{Q}_l = \dot{P}_l = -2\pi i v_l (\dot{q}_l - \dot{q}_l^*) = -4\pi^2 v_l^2 Q_l \quad \Rightarrow \quad \ddot{Q}_l + 4\pi^2 v_l^2 Q_l = 0 \qquad (9h)$$

From (8k, 9c, 9e),

$$\vec{E}(\vec{r},t) = -\frac{1}{c\sqrt{V}} \sum_l \vec{a}_l [\dot{Q}_l(t)\cos(\vec{k}_l \bullet \vec{r}) - \frac{\dot{P}_l(t)}{2\pi v_l}\sin(\vec{k}_l \bullet \vec{r})]$$

$$= -\frac{1}{c\sqrt{V}} \sum_l \vec{a}_l [P_l(t)\cos(\vec{k}_l \bullet \vec{r}) + 2\pi v_l Q_l \sin(\vec{k}_l \bullet \vec{r})] \qquad (9i)$$

$$H(\vec{r},t) = \frac{1}{\sqrt{V}} \sum_l (\vec{k}_l \wedge \vec{a}_l)[-Q_l(t)\sin(\vec{k}_l \bullet \vec{r}) - \frac{P_l}{2\pi v_l}\cos(\vec{k}_l \bullet \vec{r})]$$

$$= -\frac{1}{c\sqrt{V}} \sum_l \frac{(\vec{k}_l \wedge \vec{a}_l)}{2\pi v_l}[2\pi v_l Q_l(t)\sin(\vec{k}_l \bullet \vec{r}) + P_l(t)\cos(\vec{k}_l \bullet \vec{r})]$$

Let $(\vec{a}_l, \vec{a}_l', \vec{k}_l)$ be a right-handed set of orthogonal vectors such that,

$$|\vec{a}_l| = |\vec{a}_l'| \quad \& \quad k_l = |\vec{k}_l| \,, \quad k_l = \frac{2\pi}{\lambda_l} = \frac{2\pi}{c/\nu_l} = \frac{2\pi\nu_l}{c} \tag{9j}$$

$$\therefore \vec{H}(\vec{r},t) = -\frac{1}{c\sqrt{V}} \sum_l \vec{a}_l'[2\pi\nu_l Q_l(t)\sin(\vec{k}_l \bullet \vec{r}) + P_l(t)\cos(\vec{k}_l \bullet \vec{r})] \tag{9k}$$

Using (9i, 9k), the electromagnetic energy with the cube of side L is given by,

$$W = \frac{1}{8\pi}\int(E^2 + H^2)dV = \frac{1}{8\pi c^2}\sum_l a_l^2\left(P_l^2 + 4\pi^2\nu_l^2 Q_l^2\right) = \frac{1}{4\pi c^2}\sum_l a_l^2 H_l \tag{10}$$

Define,

$$\vec{a}_l = \sqrt{4\pi c^2}\,\vec{\varepsilon}_l \quad, \quad |\vec{\varepsilon}_l| = 1 \tag{10a}$$

From (9, 10) using (10a),

$$\vec{A}(\vec{r},t) = \frac{1}{\sqrt{V}}\sum_l \vec{a}_l[q_l(t)e^{i(\vec{k}_l\bullet\vec{r})} + q_l^*(t)e^{-i(\vec{k}_l\bullet\vec{r})}] = \frac{1}{\sqrt{V}}\sum_l [q_l(t)\vec{A}_l + q_l^*(t)\vec{A}_l^*] \tag{10b}$$

$$W = \sum_l H_l \tag{10c}$$

where,

$$\vec{A}_l = \sqrt{4\pi c^2}\,\vec{\varepsilon}_l\, e^{i\vec{k}_l\bullet\vec{r}} \tag{10d}$$

We can re-write (9f) using (9d), giving,

$$H_l = 8\pi^2\nu_l^2 q_l q_l^* \quad, \quad W = \sum_l 8\pi^2\nu_l^2 q_l q_l^* \tag{10e}$$

3. Motion of an electron : The Dirac Hamiltonian for an electron of charge e in an electromagnetic field $A(\vec{r},t), \phi(\vec{r},t))$ is

$$H = c\vec{\alpha}\bullet(\vec{p} - \frac{e}{c}\vec{A}) + \alpha_0 m_0 c^2 + e\phi \tag{11}$$

$$\therefore \quad \frac{dx}{dt} = \frac{i}{\hbar}(Hx - xH) = \frac{ic}{\hbar}\alpha_x(p_x x - xp_x) = c\alpha_x$$

or in vector notation,

$$\vec{v} = c\vec{\alpha} \tag{11a}$$

Since the eigenvalues of the $\vec{\alpha}$ operators are ± 1, the measured values of any component of velocity $= \pm c$. To understand this result consider,

$$\alpha_x \hat{H} + \hat{H}\alpha_x$$
$$= c\alpha_x(\alpha_x p_x + \alpha_y p_y + \alpha_z p_z + \alpha_0 m_0 c^2) + c(\alpha_x p_x + \alpha_y p_y + \alpha_z p_z + \alpha_0 m_0 c^2)\alpha_x$$

$$= 2cp_x \qquad (11b)$$

where in the absence of an electromagnetic field,

$$\hat{H} = c\vec{\alpha}\cdot\vec{p} + \alpha_0 m_0 c^2 \qquad (11c)$$

and, $\quad \alpha_i \alpha_j + \alpha_j \alpha_i = 2\delta_{ij} \qquad (11d)$

$\therefore$ from (11b, 11c)

$$\frac{d\alpha_x}{dt} = \frac{i}{\hbar}(\hat{H}\alpha_x - \alpha_x \hat{H}) = \frac{i}{\hbar}(2cp_x - 2\alpha_x \hat{H}) \qquad (11e)$$

But $\hat{H}, p_x$ are constants of motion, since the α's are independent of the coordinates, time, momentum, and energy.

$$\frac{d\hat{H}}{dt} = \frac{dp_x}{dt} = 0$$

integrating (11e),

$$\frac{d\alpha_x}{dt} + 2\frac{i}{\hbar}\alpha_x \hat{H} = \frac{2ic}{\hbar}p_x \qquad (11f)$$

integrating factor $= e^{\int \frac{2i\hat{H}}{\hbar}dt} = e^{\frac{2i\hat{H}}{\hbar}t}$

$\therefore \quad \dfrac{d}{dt}\left(\alpha_x e^{\frac{2i\hat{H}}{\hbar}t}\right) = \dfrac{2ic}{\hbar}p_x e^{\frac{2i\hat{H}}{\hbar}t} \quad \Rightarrow \quad \alpha_x = A'e^{-\frac{2i\hat{H}}{\hbar}t} + cp_x \hat{H}^{-1} \qquad (11g)$

From (11f, 11g),

$$\left(\frac{d\alpha_x}{dt}\right)_{t=0} = \frac{2i}{\hbar}\left[cp_x - (\alpha_x)_{t=0}\hat{H}\right] = \frac{2i}{\hbar}\left[cp_x - (A' + cp_x\hat{H}^{-1})\hat{H}\right] = -\frac{2i}{\hbar}A'\hat{H}$$

$$\therefore \quad A' = \frac{i\hbar}{2}\left(\frac{d\alpha_x}{dt}\right)_{t=0} \hat{H}^{-1} \tag{11h}$$

using (11a, 11h),

$$\frac{dx}{dt} = c^2 p_x \hat{H}^{-1} + cA'e^{-\frac{2i\hat{H}}{\hbar}t} = c^2 p_x \hat{H}^{-1} + \frac{ic\hbar}{2}\left(\frac{d\alpha_x}{dt}\right)_{t=0} \hat{H}^{-1} e^{-\frac{2i\hat{H}}{\hbar}t}$$

integrating,

$$x = c^2 p_x \hat{H}^{-1} t - \frac{1}{4}c\hbar^2 \left(\frac{d\alpha_x}{dt}\right)_{t=0} \hat{H}^{-2} e^{-\frac{2i\hat{H}}{\hbar}t} + \text{constant} \tag{11i}$$

But,

$$E = \frac{m_0 c^2}{\sqrt{1-\frac{v^2}{c^2}}}, \quad \vec{p} = \frac{m_0}{\sqrt{1-\frac{v^2}{c^2}}}\vec{v} \tag{11j}$$

Therefore the first term in (11i) is $\dot{x}t$, the classical result. The second term oscillates with a large frequency $= 2E/\hbar$.

4. The 2 x 2 Pauli spin matrices

$$\sigma_x^{(2)} = \begin{pmatrix} 0,1 \\ 1,0 \end{pmatrix}, \quad \sigma_y^{(2)} = \begin{pmatrix} 0,-i \\ i,0 \end{pmatrix}, \quad \sigma_z^{(2)} = \begin{pmatrix} 1,0 \\ 0,-1 \end{pmatrix} \tag{12}$$

satisfies the conditions,

$$\sigma_x^{(2)^2} = \sigma_y^{(2)^2} = \sigma_z^{(2)^2} = 1$$
$$\sigma_y^{(2)}\sigma_z^{(2)} + \sigma_z^{(2)}\sigma_y^{(2)} = 0$$
$$\sigma_z^{(2)}\sigma_x^{(2)} + \sigma_x^{(2)}\sigma_z^{(2)} = 0$$
$$\sigma_x^{(2)}\sigma_y^{(2)} + \sigma_y^{(2)}\sigma_x^{(2)} = 0$$

$$\sigma_y^{(2)}\sigma_z^{(2)} - \sigma_z^{(2)}\sigma_y^{(2)} = 2i\sigma_x^{(2)}$$
$$\sigma_z^{(2)}\sigma_x^{(2)} - \sigma_x^{(2)}\sigma_z^{(2)} = 2i\sigma_y^{(2)}$$
$$\sigma_x^{(2)}\sigma_y^{(2)} - \sigma_y^{(2)}\sigma_x^{(2)} = 2i\sigma_z^{(2)}$$

(12a)

In the absence of any electromagnetic fields or external force fields derivable from a potential function, the relativistic Dirac equation for an electron is,

$$\{c(\vec{\alpha}\cdot\vec{p})+\alpha_0 m_0 c^2\}\Psi = E\Psi \qquad (12b)$$

Try,

$$\Psi(\vec{r},t) = u e^{i(\vec{p}\cdot\vec{r}-Et)/\hbar} \qquad (12c)$$

$$\therefore \qquad \{c(\vec{\alpha}\cdot\vec{p})+\alpha_0 m_0 c^2\}u = Eu \qquad (12d)$$

where p, H are constants of the motion as was shown in Question 2, and u is a column vector
of size, 4×1. The $\alpha_0, \vec{\alpha}$ matrices in (12b) must satisfy,

$$\alpha_i \alpha_j + \alpha_j \alpha_i = 2\delta_{ij} \quad , \quad i,j = 0 \to 3 \qquad (12e)$$

The following 4 x 4 matrices satisfy the requirements (12d)

$$\alpha_1 = \begin{pmatrix} 0, \sigma_x^{(2)} \\ \sigma_x^{(2)}, 0 \end{pmatrix}, \quad \alpha_2 = \begin{pmatrix} 0, \sigma_y^{(2)} \\ \sigma_y^{(2)}, 0 \end{pmatrix}, \quad \alpha_3 = \begin{pmatrix} 0, \sigma_z^{(2)} \\ \sigma_z^{(2)}, 0 \end{pmatrix}, \quad \alpha_0 = \begin{pmatrix} I_{2\times 2}, 0 \\ 0, -I_{2\times 2} \end{pmatrix} \qquad (12f)$$

Let us assume without loss of generality that the electron is moving in the z- direction so that,

$$\vec{p} = (0, 0, p_z), \quad \vec{\alpha}\cdot\vec{p} = \alpha_3 p_z$$

and (12d) becomes,

$$\begin{pmatrix} m_0 c^2, 0, cp_z, 0 \\ 0, m_0 c^2, 0, -cp_z \\ cp_z, 0, -m_0 c^2, 0 \\ 0, -cp_z, 0, -m_0 c^2 \end{pmatrix} \begin{pmatrix} u_1 \\ u_2 \\ u_3 \\ u_4 \end{pmatrix} = E \begin{pmatrix} u_1 \\ u_2 \\ u_3 \\ u_4 \end{pmatrix} \qquad (12g)$$

For non-trivial solutions, the determinant,

$$\begin{vmatrix} m_0c^2 - E, 0, cp_z, 0 \\ 0, m_0c^2 - E, 0, -cp_z \\ cp_z, 0, -m_0c^2 - E, 0 \\ 0, -cp_z, 0, -m_0c^2 - E \end{vmatrix} = 0$$

This gives on simplification,

$$(m_0^2 c^4 - E^2 + c^2 p_z^2)^2 = 0 \tag{12h}$$

$$\therefore \quad E = \pm c\sqrt{m_0^2 c^2 + p_z^2} = (E_+, E_-) \tag{12i}$$

Consider the following 4 vectors,

$$\vec{U}^{(1)} = \begin{pmatrix} 1 \\ 0 \\ \dfrac{cp_z}{m_0c^2 + E_+} \\ 0 \end{pmatrix}, \vec{U}^{(2)} = \begin{pmatrix} 0 \\ 1 \\ 0 \\ -\dfrac{cp_z}{m_0c^2 + E_+} \end{pmatrix}, \vec{U}^{(3)} = \begin{pmatrix} -\dfrac{cp_z}{m_0c^2 - E_-} \\ 0 \\ 1 \\ 0 \end{pmatrix}, \vec{U}^{(4)} = \begin{pmatrix} 0 \\ \dfrac{cp_z}{m_0c^2 - E_-} \\ 0 \\ 1 \end{pmatrix}$$

$$\tag{12j}$$

We observe (12j),
- satisfies (12g)
- they are linearly independent
- satisfies (12b) → $H\vec{U}^{(1)} = E_+ \vec{U}^{(1)}$, $H\vec{U}^{(2)} = E_+ \vec{U}^{(2)}$, $H\vec{U}^{(3)} = E_- \vec{U}^{(3)}$, $H\vec{U}^{(4)} = E_- \vec{U}^{(4)}$,
- The spin angular momentum: $\vec{M}^{(s)} = \dfrac{\hbar}{2}\vec{\sigma}$, where $\vec{\sigma}$ are the 4 dimensional Dirac $\vec{\sigma}$ matrices,

$$\sigma_x = \begin{pmatrix} \sigma_x^{(2)}, 0 \\ 0, \sigma_x^{(2)} \end{pmatrix}, \quad \sigma_y = \begin{pmatrix} \sigma_y^{(2)}, 0 \\ 0, \sigma_y^{(2)} \end{pmatrix}, \quad \sigma_z = \begin{pmatrix} \sigma_z^{(2)}, 0 \\ 0, \sigma_z^{(2)} \end{pmatrix}$$

$$M_z^{(s)}\vec{U}^{(1)} = \frac{\hbar}{2}\vec{U}^{(1)}, \quad M_z^{(s)}\vec{U}^{(2)} = -\frac{\hbar}{2}\vec{U}^{(2)}, \quad M_z^{(s)}\vec{U}^{(3)} = \frac{\hbar}{2}\vec{U}^{(3)}, \quad M_z^{(s)}\vec{U}^{(4)} = -\frac{\hbar}{2}\vec{U}^{(4)}$$

$$\tag{12k}$$

To summarize the results (12k):
- $\vec{U}^{(1)}, \vec{U}^{(2)}$ correspond to positive energy eigenvalues E_+

- $\vec{U}^{(3)}, \vec{U}^{(4)}$ correspond to negative energy eigenvalues E_-
- $\vec{U}^{(1)}, \vec{U}^{(3)}$ correspond to spin up
- $\vec{U}^{(2)}, \vec{U}^{(4)}$ correspond to spin down

5. With a slight change of notation, the Hamiltonian for the pure radiation field can be written as (9f, 10c),

$$H_l = \frac{1}{2}P_l^2 + 2\pi^2 v_l^2 Q_l^2 \quad ; \quad H_f = \sum_l H_l \tag{13}$$

From (5l, 5e), the nonzero matrix elements in the Heisenberg representation are with the mass $m=1$,

$$(Q_l)_{n,n+1} = \sqrt{\frac{(n+1)\hbar}{4\pi v_l}} e^{-2\pi i v_l t} \quad ; \quad (Q_l)_{n+1,n} = \sqrt{\frac{(n+1)\hbar}{4\pi v_l}} e^{2\pi i v_l t}$$

$$(P_l)_{n,n+1} = -2\pi i v_l (Q_l)_{n,n+1} = -2\pi i v_l \sqrt{\frac{(n+1)\hbar}{4\pi v_l}} e^{-2\pi i v_l t} \tag{13a}$$

$$(P_l)_{n+1,n} = 2\pi i v_l (Q_l)_{n+1,n} = 2\pi i v_l \sqrt{\frac{(n+1)\hbar}{4\pi v_l}} e^{2\pi i v_l t}$$

where the self-adjoint operators P_l, Q_l satisfy the usual commutation relations,

$$Q_l Q_m - Q_m Q_l = 0, \quad P_l P_m - P_m P_l = 0, \quad Q_l P_m - P_m Q_l = i\hbar \delta_{lm} \tag{13b}$$

$$\frac{dQ_l}{dt} = \frac{i}{\hbar}[H_f Q_l - Q_l H_f] \quad , \quad \frac{dP_l}{dt} = \frac{i}{\hbar}[H_f P_l - P_l H_f] \tag{13c}$$

Define the self-adjoint operators,

$$Q_l = q_l + q_l^+ \quad , \quad P_l = -2\pi i v_l (q_l - q_l^+) \tag{13d}$$

Solving for q_l, q_l^+

$$q_l = \frac{1}{2}\left(Q_l + \frac{i}{2\pi v_l}P_l\right) \quad , \quad q_l^+ = \frac{1}{2}\left(Q_l - \frac{i}{2\pi v_l}P_l\right) \tag{13e}$$

using (13a) in (13e), the only nonzero matrix representatives of q_l, q_l^+ are,

$$(q_l)_{n,n+1} = \sqrt{\frac{(n+1)\hbar}{4\pi v_l}} e^{-2\pi i v_l t} \quad ; \quad (q_l^+)_{n+1,n} = \sqrt{\frac{(n+1)\hbar}{4\pi v_l}} e^{2\pi i v_l t} \tag{13f}$$

The energy eigenvalues are given by (10e, 13f),

$$(E_l)_n = (H_l)_{nn} = 8\pi^2 v_l^2 (q_l^+ q_l)_{nn} = 8\pi^2 v_l^2 (q_l^+)_{n,n-1}(q_l)_{n-1,n} = 8\pi^2 v_l^2 \frac{n\hbar}{4\pi v_l} = nhv_l \tag{13g}$$

$(q_l)_{n,n+1} \equiv$ transition from a state with energy $(n+1)hv_l$ to a state with energy nhv_l. This implies $q_l \to$ operator for the **emission** of a photon of frequency v_l by the *l*-th oscillator.

$(q^+_l)_{n,n+1} \equiv$ transition from a state with energy nhv_l to a state with energy $(n+1)hv_l$. This implies $q_l^+ \to$ operator for the **absorption** of a photon of frequency v_l by the *l*-th oscillator.

The quantum mechanical analogue of (10b, 10d) is,

$$\vec{A}(\vec{r},t) = \sqrt{\frac{4\pi c^2}{V}} \sum_l \vec{\varepsilon}_l [q_l(t) e^{i\vec{k}_l \cdot \vec{r}} + q_l^+(t) e^{-i\vec{k}_l \cdot \vec{r}}] \tag{13h}$$

with the emission and absorption operator given by (13f). From (11) we see that the interaction Hamiltonian for a system of electrons is given by,

$$H_{int} = -e \sum_s \vec{\alpha}_s \cdot \vec{A}^{(s)} \tag{13i}$$

For a single electron,
$$H_{int} = -e\vec{\alpha} \cdot \vec{A} \tag{13j}$$

where,

Let, ψ_a = normalized time independent wave function of electron before emission of a photon.

ψ_b = normalized time independent wave function of electron after emission of a photon.

ψ_0 = normalized time independent wave function of oscillator before absorption of the photon.

ψ_1 = normalized time independent wave function of oscillator after absorption of the photon.

q_l, q_l^+ are respectively the matrix operators for the **emission** and **absorption** of a photon of frequency v_l by the l-th harmonic oscillator. For emission and absorption of a photon by an electron, the role of q_l, q_l^+ are reversed. For example the vector potential for the emission of a photon by an electron is given by (13h),

$$\vec{A}(\vec{r},t) = \sqrt{\frac{4\pi c^2}{V}} \vec{\varepsilon} q^+(t) e^{-i\vec{k}\cdot\vec{r}} \tag{13k}$$

∴ from (13j),

$$H_{int} = -e\sqrt{\frac{4\pi c^2}{V}} (\vec{\alpha}\cdot\vec{\varepsilon}) q^+(t) e^{-i\vec{k}\cdot\vec{r}} \tag{13l}$$

where $\vec{r}$ is evaluated at the site of the electron. Remember: emission / absorption by an electron ≡ absorption / emission by an oscillator.

initial wave function of the system consisting of electron + oscillator = $\psi_a \psi_0$
final wave function of the system consisting of electron + oscillator = $\psi_b \psi_1$

∴ the interaction matrix element representing the emission of the photon by the electron and its subsequent absorption by the oscillator is ,

$$H_{b,1;a,0}^{int} = \int \psi_b^* \psi_1^* H_{int} \psi_a \psi_0$$

From (13l),

$$H_{b,1;a,0}^{int} = -ec\sqrt{\frac{4\pi}{V}} \int \psi_b^* \psi_1^* (\vec{\alpha}\cdot\vec{\varepsilon}) q^+(t) e^{-i\vec{k}\cdot\vec{r}} \psi_a \psi_0 \tag{13m}$$

The integration is over the position vector $\vec{r}$ of the electron, and the coordinates of the oscillator. Integration with respect to the oscillator gives from (13f),

$$\int \psi_1^* q^+(t) \psi_0 = q_{10}^+ = \sqrt{\frac{\hbar}{4\pi\nu}} e^{2\pi i \nu t} \tag{13n}$$

From (12c) for free electron,

$$\psi_a(\vec{r},t) = u_a e^{i(\vec{p}_a\cdot\vec{r}-E_a t)/\hbar} \; ; \qquad \psi_b(\vec{r},t) = u_b e^{i(\vec{p}_b\cdot\vec{r}-E_b t)/\hbar} \tag{13o}$$

Substituting (13n, 13o) in (13m)

$$H_{b,1;a,0}^{int} = -ec\sqrt{\frac{\hbar}{V\nu}} \int u_b^+ (\vec{\alpha}\cdot\vec{\varepsilon}) e^{\frac{i}{\hbar}(\vec{p}_a-\vec{p}_b-\hbar\vec{k})\cdot\vec{r}} u_a d^3\vec{r} e^{i(2\pi\nu+\frac{E_b}{\hbar}-\frac{E_a}{\hbar})t}$$

$$= -ec\sqrt{\frac{\hbar}{V\nu}} e^{i(2\pi\nu+\frac{E_b}{\hbar}-\frac{E_a}{\hbar})t} u_b^+ (\vec{\alpha}\cdot\vec{\varepsilon}) u_a \int e^{\frac{i}{\hbar}(\vec{p}_a-\vec{p}_b-\hbar\vec{k})\cdot\vec{r}} d^3\vec{r}$$

$$= -ec(2\pi\hbar)^3 \sqrt{\frac{\hbar}{Vv}} e^{i(2\pi v + \frac{E_b}{\hbar} - \frac{E_a}{\hbar})t} u_b^+ (\vec{\alpha} \bullet \vec{\varepsilon}) u_a \delta(\vec{p}_a - \vec{p}_b - \hbar\vec{k})$$

$$= -ech^3 \sqrt{\frac{\hbar}{Vv}} e^{i(2\pi v + \frac{E_b}{\hbar} - \frac{E_a}{\hbar})t} u_b^+ (\vec{\alpha} \bullet \vec{\varepsilon}) u_a \tag{13p}$$

The Dirac delta function is zero unless,

$$\vec{p}_a = \vec{p}_b + \hbar\vec{k} \rightarrow \text{conservation of linear momentum}$$

6. For a perfect gas consisting of N atoms in a container of volume V at a temperature T, the partition function is given by,

$$Z = \int e^{-\frac{H}{kT}} d\lambda \quad ; \quad H = \sum_{i=1}^{N} \frac{p_i^2}{2m} \quad ; \quad d\lambda = \prod_{i=1}^{N} d\vec{q}_i d\vec{p}_i \quad ; \quad d\vec{q}_i = dx_i dy_i dz_i \quad ;$$
$$d\vec{p}_i = dp_{xi} dp_{yi} dp_{zi} \tag{14}$$

Therefore,

$$Z = \prod_{i=1}^{N} \int_{-\infty}^{\infty} e^{-\frac{p_i^2}{2m}} d\vec{p} d\vec{q}_i = \left(V \int_{-\infty}^{\infty} e^{-\frac{p_i^2}{2m}} dp_i \right)^N = \left[V(2\pi mkT)^{\frac{3}{2}} \right]^N \tag{14a}$$

$$\therefore \quad \ln Z = N \left[\ln V + \frac{3}{2} (\ln(2\pi m) - \ln \beta) \right] \quad ; \quad \beta = \frac{1}{kT} \quad ; \quad k = \text{Boltzmann's constant,}$$

From statistical mechanics,

$$E = -\frac{\partial \ln Z}{\partial \beta} = \frac{3N}{2\beta} = \frac{3NkT}{2} \tag{14b}$$

The specific heat at constant volume,

$$C_V = \frac{\partial E}{\partial T} = \frac{3Nk}{2} \tag{14c}$$

The entropy is given by,

$$S = k \left(\ln Z - \beta \frac{\partial \ln Z}{\partial \beta} \right) = Nk \left(\ln V - \frac{3}{2} \ln \beta + cons \tan t \right) \tag{14d}$$

173

and the free energy,

$$F = E - TS = -\frac{1}{\beta}\ln Z \qquad (14e)$$

$$\therefore \quad dF = dE - TdS - SdT = dE - \delta Q - SdT = -\delta W - SdT = -pdV - SdT \qquad (14f)$$

where δW = work done by the system. At constant temperature $T = constant$.

$$dF = -pdV \quad \Rightarrow \quad -\frac{1}{\beta}d\ln Z = -pdV \quad \Rightarrow \quad -\frac{1}{\beta}\frac{\partial \ln Z}{\partial V}dV = -pdV$$

$$\therefore \quad p = \frac{1}{\beta}\frac{\partial \ln Z}{\partial V} = \frac{N}{\beta V} = \frac{NkT}{V} \qquad (14g)$$

The probability density for the canonical ensemble is,

$$\rho = \frac{e^{-\beta H}}{Z} \qquad (14h)$$

$\therefore$ the probability that $\vec{p}_1 = \vec{p}$, $\vec{r}_1 = \vec{r}$ in a small volume $d\vec{q}_1 d\vec{p}_1 = dxdydzdp_x dp_y dp_z$ of phase space is,

$$P(\vec{r},\vec{p})dxdydzdp_x dp_y dp_z = \frac{1}{Z}\int dx_1 dy_1 dz_1 dp_{x1} dp_{y1} dp_{z1} e^{-\frac{\beta p_1^2}{2m}}\prod_{i=2}^{N}\left(\int_{-\infty}^{\infty} e^{-\frac{\beta p_i^2}{2m}}d\vec{q}_i d\vec{p}_i\right)$$

$$= \frac{dxdydzdp_x dp_y dp_z}{Z} e^{-\frac{\beta p^2}{2m}}\prod_{i=2}^{N}\left(\int_{-\infty}^{\infty} e^{-\frac{\beta p_i^2}{2m}}d\vec{q}_i d\vec{p}_i\right)$$

$$= \frac{dxdydzdp_x dp_y dp_z}{V(2\pi mkT)^{\frac{3}{2}}} e^{-\frac{p^2}{2mkT}}$$

$\therefore$ the number of atoms in the gas with momenta in the range $(\vec{p}, \vec{p}+d\vec{p})$, and position vectors in the range $(\vec{r}, \vec{r}+d\vec{r})$ is given by,

$$P(\vec{r},\vec{p})dxdydzdp_x dp_y dp_z = \frac{f(\vec{r},\vec{p})dxdydzdp_x dp_y dp_z}{N}$$

$$f(\vec{r},\vec{p}) = \frac{Ne^{-\frac{p^2}{2mkT}}}{V(2\pi mkT)^{\frac{3}{2}}} \qquad (14i)$$

7. **Virial Theorem**: Consider a classical dynamical system consisting of N particles with each with 3 degrees of

$$(p_1, p_2, ..., p_n) = \text{generalized momenta} \equiv (\xi_1, \xi_2, ..., \xi_n)$$

$$(q_1, q_2, ..., q_n) = \text{generalized coordinates} \equiv (\xi_{n+1}, \xi_{n+2}, ..., \xi_{2n})$$

$$n = 3N$$

and the Hamiltonian,

$$H(p_1 p_2, ..., p_n; q_1, q_2, ..., q_n) = H(\xi_1, \xi_2, ..., \xi_n; \xi_{n+1}, \xi_{n+2}, ..., \xi_{2n}) \qquad (15)$$

The element of volume in phase space is,

$$d\lambda = \prod_{i=1}^{n} dp_i dq_i = \prod_{k=1}^{2n} d\xi_k \qquad (15a)$$

Consider the mean value of (no summation over repeated suffix),

$$\langle \xi_k \frac{\partial H}{\partial \xi_k} \rangle = \int \xi_k \frac{\partial H}{\partial \xi_k} \rho(\xi_1, ..., \xi_{2n}) d\lambda = \frac{\int \xi_k \frac{\partial H}{\partial \xi_k} e^{-\beta H} d\lambda}{\int e^{-\beta H} d\lambda} \quad ; \quad \beta = \frac{1}{kT} \qquad (15b)$$

where for a canonical ensemble the probability density is given by (14, 14h). Therefore, using the identity,

$$e^{-\beta H} \frac{\partial H}{\partial \xi_k} = -\frac{1}{\beta} \frac{\partial}{\partial \xi_k} e^{-\beta H}$$

$$\langle \xi_k \frac{\partial H}{\partial \xi_k} \rangle = -\frac{\int \xi_k \frac{\partial e^{-\beta H}}{\partial \xi_k} d\lambda}{\beta \int e^{-\beta H} d\lambda} = \frac{\int e^{-\beta H} d\lambda}{\beta \int e^{-\beta H} d\lambda} = \frac{1}{\beta} = kT \qquad (15c)$$

using integration by parts and appropriate boundary conditions.

Van der Waal's equation : In the presence of a potential function,

$$H = \sum_{s=1}^{N} \frac{\vec{p}_s^2}{2m_s} + U(\vec{r}_1, \vec{r}_2, ..., \vec{r}_N) \tag{15d}$$

Case (1): $\xi_k = p_{sx}$ (the x-component of the momentum of the sth particle) →

$$\frac{\partial H}{\partial \xi_k} = \frac{p_{sx}}{m_s} \text{ from (15d)}$$

Therefore, the Virial Theorem (15c) gives,

$$\langle \xi_k \frac{\partial H}{\partial \xi_k} \rangle = \langle \frac{p_{sx}^2}{m_s} \rangle = kT \quad \Rightarrow \quad \langle \frac{\vec{p}_s^2}{m_s} \rangle = 3kT$$

Using (15d),

$$\langle H \rangle = \frac{3}{2} NkT + \langle U \rangle \tag{15e}$$

Case (2): $\xi_k = x_s$ (the x-component of the position vector of the sth particle) →

$$\frac{\partial H}{\partial \xi_k} = \frac{\partial U}{\partial x_s} = -F_{sx} \text{ from (15d)}$$

Therefore, the Virial Theorem (15c) gives,

$$\langle \xi_k \frac{\partial H}{\partial \xi_k} \rangle = \langle -x_s F_{sx} \rangle = kT \quad \Rightarrow \quad -\langle \vec{r}_s \bullet \vec{F}_s \rangle = 3kT \tag{15f}$$

Therefore,

$$\Xi_1 \equiv -\sum_{s=1}^{N} \langle \vec{r}_s \bullet \vec{F}_s \rangle = 3NkT \tag{15g}$$

Define,

$\vec{F}_{ss'}$ = force acting on the particle s due to particle s'

Then the total force acting on particle s due to all the other particles is,

$$\vec{F}_s^T = \sum_{s'=1}^{N} \vec{F}_{ss'} \quad ; \quad \vec{F}_{ss} = \vec{0} \tag{15h}$$

then,

$$\Xi_2 \equiv -\sum_{s=1}^{N}\langle \vec{r}_s \bullet \vec{F}_s^T \rangle = -\sum_{s=1}^{N}\sum_{s'=1}^{N}\langle \vec{r}_s \bullet \vec{F}_{ss'} \rangle \tag{15i}$$

But by Newton's third law of motion,

$$\vec{F}_{ss'} = -\vec{F}_{s's} \tag{15j}$$

$$\therefore \Xi_2 \equiv -\sum_{s=1}^{N}\sum_{s'=1}^{N}\langle \vec{r}_s \bullet \vec{F}_{ss'} \rangle = \sum_{s=1}^{N}\sum_{s'=1}^{N}\langle \vec{r}_s \bullet \vec{F}_{s's} \rangle = \sum_{s=1}^{N}\sum_{s'=1}^{N}\langle \vec{r}_{s'} \bullet \vec{F}_{ss'} \rangle$$

interchanging the dummy suffices s, s',

$$\therefore 2\Xi_2 = -\sum_{s=1}^{N}\sum_{s'=1}^{N}\langle (\vec{r}_s - \vec{r}_{s'}) \bullet \vec{F}_{ss'} \rangle \tag{15k}$$

$$\therefore \quad \Xi_2 = -\frac{1}{2}\sum_{s=1}^{N}\sum_{s'=1}^{N}\langle \vec{r}_{ss'} \bullet \vec{F}_{ss'} \rangle \tag{15l}$$

where,

$$\vec{r}_{ss'} = \vec{r}_s - \vec{r}_{s'} \tag{15m}$$

We shall assume that the interaction potential depends only on the magnitude of the distance between the particles,

$$U(|\vec{r}_s - \vec{r}_{s'}|) = U(r_{ss'}) \tag{15n}$$

Then,

$$\vec{F}_{ss'} = -\nabla U(r_{ss'})$$

and,

$$\langle \vec{r}_{ss'} \bullet \vec{\nabla}_s U(r_{ss'}) \rangle = \langle \vec{r}_{s's} \bullet \vec{\nabla}_{s'} U(r_{ss'}) \rangle \rightarrow \text{symmetric in } s, s'. \tag{15o}$$

For identical particles,

$$\langle \vec{r}_{ss'} \bullet \vec{\nabla}_s U(r_{ss'}) \rangle \equiv \langle \vec{r} \bullet \vec{\nabla} U(r) \rangle$$

and from (15l),

$$\Xi_2 = \frac{N(N-1)}{2}\langle \vec{r} \bullet \vec{\nabla} U(r) \rangle \cong \frac{N^2}{2}\langle \vec{r} \bullet \vec{\nabla} U(r) \rangle \tag{15p}$$

But,

$$\langle \vec{r} \bullet \vec{\nabla} U(r) \rangle = \frac{\int \vec{r}_{ss'} \bullet \vec{\nabla} U(r_{ss'}) e^{-\beta H} d\lambda}{\int e^{-\beta H} d\lambda} = \frac{\int \vec{r}_{ss'} \bullet \vec{\nabla} U(r_{ss'}) e^{-\beta U(r_{ss'})} d\vec{r}_s d\vec{r}_{s'}}{\int e^{-\beta U(r_{ss'})} d\vec{r}_s d\vec{r}_{s'}} \quad (15q)$$

Changing to a center of mass coordinate system,

$$\vec{r} = \vec{r}_s - \vec{r}_{s'} = \vec{r}_{ss'} \quad ; \quad \vec{R} = (\vec{r}_s + \vec{r}_{s'})/2$$

Therefore,

$$\vec{r}_s = \vec{R} + \frac{1}{2}\vec{r} \quad , \quad \vec{r}_{s'} = \vec{R} - \frac{1}{2}\vec{r} ,$$

The Jacobian of the transformation is,

$$d\vec{r}_s d\vec{r}_{s'} = \left| \frac{\partial(\vec{r}_s, \vec{r}_{s'})}{\partial(\vec{r}, \vec{R})} \right| d\vec{r} d\vec{R} = \det \begin{vmatrix} \frac{1}{2}\vec{I}, \vec{I} \\ -\frac{1}{2}\vec{I}, \vec{I} \end{vmatrix} d\vec{r} d\vec{R} = d\vec{r} d\vec{R}$$

where, $\vec{I}$ is the 3 x 3 unit matrix. Therefore (15q) reduces to,

$$\langle \vec{r} \bullet \vec{\nabla} U(r) \rangle = \frac{\int \vec{r} \bullet \vec{\nabla} U(r) e^{-\beta U(r)} d\vec{r}}{\int e^{-\beta U(r)} d\vec{r}} \quad (15r)$$

Therefore the effective probability density is

$$\rho^* = \frac{e^{-\beta U(r)}}{\int e^{-\beta U(r)} d\vec{r}} \quad (15s)$$

Since, the region over which $U(r) \neq 0$, is very small, we can approximate,

$$\int e^{-\beta U(r)} d\vec{r} \cong V \text{, the volume of the container.}$$

$$\therefore \quad \langle \vec{r} \bullet \vec{\nabla} U(r) \rangle = \langle r \frac{dU}{dr} \rangle = \frac{1}{V} \int r \frac{dU}{dr} e^{-\beta U} 4\pi r^2 dr = \frac{4\pi}{V} \int_0^\infty r^3 \frac{dU}{dr} e^{-\beta U} dr \quad (15s)$$

From (15p),

$$\Xi_2 \cong \frac{2\pi N^2}{V} \int_0^\infty r^3 \frac{dU}{dr} e^{-\beta U} dr \tag{15t}$$

using the identity,

$$\frac{dU}{dr} e^{-\beta U} = -\frac{1}{\beta} \frac{de^{-\beta U}}{dr} = -\frac{1}{\beta} \frac{d(e^{-\beta U} - 1)}{dr}$$

we can simplify (15t) by integrating by parts,

$$\Xi_2 = -\frac{2\pi N^2}{\beta V} \int_0^\infty r^3 \frac{d}{dr}(e^{-\beta U} - 1) dr = \frac{2\pi N^2}{\beta V} \int_0^\infty 3r^2 (e^{-\beta U} - 1) dr \tag{15u}$$

assuming,

$$[r^3(e^{-\beta U} - 1)]_0^\infty = 0 \;\; \rightarrow \;\; U(r) \text{ goes to zero sufficiently fast as } r \text{ tends to infinity.}$$

The virial for a system of N particles in a cube of side L, can be calculated as follows: when a particle strikes a wall of the container, it experiences an inward force directing it towards the interior of the box. For purposes of simplicity, consider 2 parallel faces of the cube at $x=0$ & $x=L$. The inward force pushing a particle that strikes the wall is,

$$F_x = -\frac{\partial U_w}{\partial x} \tag{15v}$$

where U_w is the "wall potential" approximated by a square well of side L,

$$U_w = 0 \; , \; \text{inside box}$$
$$ = \infty \; , \; \text{outside box} \tag{15w}$$

using (15g), for the x-component of the "wall force"

$$\Xi_x = -\sum_{s=1}^{N} \langle x_s F_{sx} \rangle \Big|_{x=0}^{x=L} = -L \sum_{s=1}^{N} \langle F_{sx} \rangle \Big|_{x=L} \tag{15x}$$

But, from Newton's third law, the force exerted by the particles on the wall must be equal and opposite to the force exerted by the wall on the particles,

$$\therefore \;\; -\sum_{s=1}^{N} \langle F_{sx} \rangle \Big|_{x=L} = F_W = \text{force exerted by the particles on the wall}$$
$$= \text{(gas) pressure} \times \text{area of wall} = pL^2 \tag{15y}$$

From (15x),

$$\Xi_x = pL^3 = pV$$

Similarly,
$$\Xi_y = \Xi_z = pV$$

$$\therefore \Xi_1 = -\langle \sum_{s=1}^{N} \vec{r}_s \bullet \vec{F}_{Ws} \rangle = 3pV \tag{15z}$$

Applying the virial theorem (15g), using (15u ,15z)

$$\vec{F}_s = \vec{F}_{Ws} + \vec{F}_{Us}$$

where $\vec{F}_{Us}$ = interaction forces between pairs of atoms. Therefore,

$$\therefore -\langle \sum_{s=1}^{N} \vec{r}_s \bullet \vec{F}_{Ws} \rangle \quad -\langle \sum_{s=1}^{N} \vec{r}_s \bullet \vec{F}_{Us} \rangle = 3NkT$$

$$\Xi_1 + \Xi_2 = 3NkT \tag{16}$$

$$\therefore 3pV + \frac{2\pi N^2}{\beta V} \int_0^\infty 3r^2 (e^{-\beta U} - 1)dr = 3NkT$$

$$pV = NkT - \frac{2\pi N^2}{\beta V} \int_0^\infty r^2 (e^{-\beta U} - 1)dr = NkT\left(1 - \frac{2\pi N^2}{NkT\beta V} \int_0^\infty r^2 (e^{-\beta U} - 1)dr\right)$$

$$= NkT\left(1 - \frac{b_2}{V}\right) \tag{16a}$$

where,

$$b_2 = 2\pi N \int_0^\infty r^2 (e^{-\beta U} - 1)dr \, , \quad \because \beta = \frac{1}{kT} \tag{16b}$$

Let us reconcile these relationships with the more usual form of Van der Waals equation,

$$\left(p + \frac{a}{V^2}\right)(V - b) = NkT \tag{16c}$$

Therefore using the binomial theorem for a negative integral index,

$$pV = \frac{NkT}{1-\frac{b}{V}} - \frac{a}{V} = NkT\left(1 + \frac{b}{V} - \frac{a}{VNkT}\right) \qquad (16d)$$

For a Van der Waals potential function defined by,

$$U(r) = \infty \qquad 0 \leq r \leq r_1$$
$$|U(r)| \ll kT \qquad r_1 < r \leq \infty \quad \Rightarrow \quad \beta U \ll 1 \qquad (16e)$$

then

$$e^{-\beta U} - 1 = -1 \,, \qquad 0 \leq r \leq r_1$$
$$= -\beta U \,, \qquad r_1 < r \leq \infty \qquad (16f)$$

and from (16b),

$$b_2 = 2\pi N \int_0^\infty r^2 (e^{-\beta U} - 1) dr \cong -2\pi N \int_0^{r_1} r^2 dr - \frac{2\pi N}{kT} \int_{r_1}^\infty r^2 U(r) dr \qquad (16g)$$

and equation (16a) becomes,

$$pV = NkT\left(1 + \frac{2\pi N r_1^3}{3V} + \frac{2\pi N}{kTV} \int_{r_1}^\infty r^2 U(r) dr\right) \qquad (16h)$$

Comparing (16h) with (16) we have (16d),

$$b = \frac{2\pi N r_1^3}{3} \,, \qquad a = -2\pi N^2 \int_{r_1}^\infty r^2 U(r) dr \qquad (16i)$$

8. We use the method of Lagrange multipliers, and seek the unconstrained maximum of

$$f(a_1, a_2, a_3, \ldots, a_l, \ldots) = \log P - \lambda \sum_l a_l - \beta \sum_l a_l \varepsilon_l \qquad (17)$$

where λ, β are Lagrange multipliers, and for $N \to \infty$, we use Stirling's formula
$$\log(n!) = n(\log n - 1) \qquad (17a)$$

$$\therefore \quad f(a_1, a_2, a_3, \ldots, a_l, \ldots) = \log N! - \sum_l a_l (\log a_l - 1 + \lambda + \beta \varepsilon_l) \qquad (17b)$$

$$\therefore \quad \frac{\partial f}{\partial a_l} = 0, \quad l = 1,2,3,\ldots \Rightarrow \log a_l + \lambda + \beta\varepsilon_l - 1 + 1 = 0$$

$$\therefore \quad a_l = e^{-\lambda - \beta\varepsilon_l} \tag{17c}$$

$$\therefore \quad \sum_l a_l = N \Rightarrow a_l = N \frac{e^{-\beta\varepsilon_l}}{\sum_l e^{-\beta\varepsilon_l}} = -\frac{N}{\beta} \frac{\partial}{\partial \varepsilon_l} \log \sum_l e^{-\beta\varepsilon_l} \tag{17d}$$

The average energy of the system is given by,

$$U = \frac{E}{N} = \frac{1}{N} \sum_l \varepsilon_l a_l = \frac{\sum_l \varepsilon_l e^{-\beta\varepsilon_l}}{\sum_l e^{-\beta\varepsilon_l}} = -\frac{\partial}{\partial \beta} \log \sum_l e^{-\beta\varepsilon_l} \tag{17e}$$

Interpretation of β :

$$\text{Define,} \quad F(\varepsilon_1, \varepsilon_2, \varepsilon_3, \ldots, \varepsilon_l, \ldots; \beta) = \log \sum_l e^{-\beta\varepsilon_l} \tag{18}$$

$\therefore$ from (17d, 17e) equation (18) becomes,

$$dF = \sum_l \frac{\partial F}{\partial \varepsilon_l} d\varepsilon_l + \frac{\partial F}{\partial \beta} d\beta = -\frac{\beta}{N} \sum_l a_l d\varepsilon_l - U d\beta$$

$$\therefore \quad d(F + U\beta) = \beta \left(dU - \frac{1}{N} \sum_l a_l d\varepsilon_l \right) \tag{18a}$$

But $d\varepsilon_l$ = increase in the energy of each of the a_l systems. So that $a_l d\varepsilon_l$ = increase in the energy of the systems in the l th state. This expression constitutes the work done on the a_l systems. Therefore, the work done by the a_l assemblages that make up the N identical systems discussed above $= -a_l d\varepsilon_l$. Hence the average work done by each of the systems $= -\frac{1}{N} \sum_l a_l d\varepsilon_l$. Therefore, it is highly suggestive that the terms inside the brackets in (18a) correspond to the heat δQ supplied, and $F + U\beta$ is the entropy.

Let, $\quad G = F + U\beta \tag{18b}$

$$\frac{1}{T\beta} = \phi(G) \tag{18c}$$

From (18a),

$$\phi(G)dG = \frac{\delta Q}{T} = dS \quad , \quad S = \text{entropy} \tag{18d}$$

Integrating, yields S as a function of G, or conversely,

$$G = \chi(S) \tag{18e}$$

using (17d, 17e, 18) in (18b) gives,

$$G = \log \sum_l e^{-\beta \varepsilon_l} - \beta \frac{\partial}{\partial \beta} \log \sum_l e^{-\beta \varepsilon_l} \tag{18f}$$

Consider 3 different assemblies of systems → A, B, $A+B$, with energy levels respectively, $\alpha_k, \beta_m, \varepsilon_l = \alpha_k + \beta_m$.

$$\therefore \sum_l e^{-\beta \varepsilon_l} = \sum_k \sum_m e^{-\beta(\alpha_k + \beta_m)} = \left(\sum_k e^{-\beta \alpha_k}\right)\left(\sum_m e^{-\beta \beta_m}\right) \tag{18g}$$

Let, $\chi_A(S_A), \chi_B(S_B), \chi_{A+B}(S_{A+B})$ be the corresponding χ - functions (18e) for the 3 assemblages A, B, $A+B$. Using (18g) in (18f) gives,

$$G_A + G_B = \chi_A(S_A) + \chi_B(S_B) = G_{A+B} = \chi_{A+B}(S_{A+B}) \tag{18h}$$

But, the entropy is also additive,

$$S_{A+B} = S_A + S_B + C \tag{18i}$$

where C is independent of S_A, S_B.

$$\therefore \quad \chi_A(S_A) + \chi_B(S_B) = \chi_{A+B}(S_A + S_B + C) \tag{18j}$$

differentiating (18j) with respect to S_A, followed by S_B we have

$$\chi'_A(S_A) = \chi'_{A+B}(S_{A+B}) = \chi'_B(S_B) \tag{18k}$$

$$\therefore \quad \chi'_A(S_A) = \chi'_B(S_B) = \frac{1}{k} \quad , \qquad k = \text{universal constant} \tag{18l}$$

using (18l) in (18e, 18c, 18d),

$$\frac{1}{kT\beta}dS = dS \quad \Rightarrow \quad \beta = \frac{1}{kT} \tag{18m}$$

9. Consider the integral,

$$I(n) = \int_1^n \log x \, dx, \quad n = \text{positive integer} \tag{19}$$

$I(n)$ is a monotonic increasing function of $n \geq 1$.

$$I(n) = \sum_{k=1}^{n-1} \int_k^{k+1} \log x \, dx$$

$$\therefore \quad \int_k^{k+1} \log(k) \, dx \leq \int_k^{k+1} \log x \, dx \leq \int_k^{k+1} \log(k+1) \, dx$$

Therefore,

$$\sum_{k=1}^{n-1} \int_k^{k+1} \log(k) \, dx \leq \sum_{k=1}^{n-1} \int_k^{k+1} \log x \, dx \leq \sum_{k=1}^{n-1} \int_k^{k+1} \log(k+1) \, dx$$

i.e,

$$\sum_{k=1}^{n-1} \log(k) \leq I(n) \leq \sum_{k=1}^{n-1} \log(k+1)$$

$$\log 1 + \log 2 + \log 3 + \ldots + \log(n-1) \leq \int_1^n \log x \, dx \leq \log 2 + \log 3 + \log 4 + \ldots + \log n$$

But, $\int_1^n \log x \, dx = n(\log n - 1) - 1$, integration by parts

$$\therefore \quad \log(n-1)! \leq n(\log n - 1) - 1 \leq \log n! \tag{19a}$$

as $n \to \infty$, he above inequality collapses into

$$\log n! = n(\log n - 1) \tag{19b}$$

From (18e, 18l),

$$\frac{dG}{dS} = \chi'(S) = \frac{1}{k}$$

$$\therefore \quad S = kG + \text{constant} = k \log \sum_l e^{-\beta \varepsilon_l} + \frac{U}{T} + \text{constant} \tag{20}$$

184

Equation (20) follows from (18, 18b, 18m).

Nernst Theorem: Define a partition function,

$$Z = \sum_l e^{-\beta \varepsilon_l} = \sum_l e^{-\frac{\varepsilon_l}{kT}} = \sum_l g_l e^{-\frac{\varepsilon_l}{kT}} \qquad (20a)$$

where a degeneracy g_l is assumed. Let ε_1 be the lowest energy eigenvalue and ε_2 the next lowest value. For, $kT \ll \varepsilon_2 - \varepsilon_1$,

$$Z = g_1 e^{-\frac{\varepsilon_1}{kT}} + g_2 e^{-\frac{\varepsilon_2}{kT}} = g_1 e^{-\frac{\varepsilon_1}{kT}} \left(1 + \frac{g_2}{g_1} e^{-\frac{\varepsilon_2 - \varepsilon_1}{kT}} + ..\right) \qquad (20b)$$

$$\therefore \log Z = \log g_1 - \frac{\varepsilon_1}{kT} + \log\left(1 + \frac{g_2}{g_1} e^{-\frac{\varepsilon_2-\varepsilon_1}{kT}} + ..\right) = \log g_1 - \frac{\varepsilon_1}{kT} + \frac{g_2}{g_1} e^{-\frac{\varepsilon_2-\varepsilon_1}{kT}} \qquad (20c)$$

The issue is the ramifications of taking the integration of constant in equation (20) to be zero.
From (14e) the free energy is,

$$F = U - TS$$
$$\therefore dF = dU - TdS - SdT = dU - \delta Q - SdT = -pdV - SdT$$
$$\therefore S = -\frac{\partial F}{\partial T} \qquad (20d)$$

and from (20), $\quad F = -kT \log Z \qquad (20e)$

$$= -kT \left(\log g_1 - \frac{\varepsilon_1}{kT} + \frac{g_2}{g_1} e^{-\frac{\varepsilon_2-\varepsilon_1}{kT}}\right)$$

$$= \varepsilon_1 - kT \log g_1 - kT \frac{g_2}{g_1} e^{-\frac{\varepsilon_2-\varepsilon_1}{kT}} \qquad (20f)$$

Therefore from (20d), the entropy is given by,

$$S = k \log g_1 + \left(k + \frac{\varepsilon_2 - \varepsilon_1}{T}\right) e^{-\frac{\varepsilon_2-\varepsilon_1}{kT}} \qquad (20g)$$

$$\therefore \underset{T \to 0}{Lt} S = k \log g_1 \qquad (20h)$$

10. (a) *Avogadro's Number* : Consider an element of the periodic table with atomic weight A and atomic number P. Suppose, in addition the atom contains N neutrons.

Then neglecting the mass of the electrons in neutral atom, the mass of an atom is given by,

$$M = Pm_p + Nm_n \qquad (21)$$

where,
$$m_p = \text{mass of a proton}$$
$$m_n = \text{mass of a neutron}$$

but, $\quad m_p \cong m_n = m$

$$\therefore \quad M = (P+N)m \qquad \text{gms}$$

and, the mass of a hydrogen atom $= m$

Mass of an atom of the element $= A$

Mass of an element of the hydrogen atom

$$\therefore \quad (P+N)m = Am \quad \Rightarrow \quad A = P+N$$

M gms $\equiv 1$ atom

$$\therefore \; 1 \text{ gm} \equiv \frac{1}{M} = \frac{1}{(P+N)m} = \frac{1}{Am} \quad \text{atoms}$$

$$\therefore \; A \text{ gm} \equiv \frac{A}{Am} = \frac{1}{m} \quad \text{atoms}$$

but, the mass of a proton, $m = 1.67 \times 10^{-24}$ gms

Therefore, Avogadro's Number, which corresponds to the number of atoms in A gms of the element $= \dfrac{1}{1.67 \times 10^{-24}} = 6 \times 10^{23}$

(b) Suppose for some positive integer N,

$$(x_1 + x_2 + x_3 + ... + x_l + ...)^N = \sum_{a_l} \frac{N!}{a_1! a_2! a_3! ... a_l! ...} x_1^{a_1} x_2^{a_2} x_3^{a_3} .... x_l^{a_l} ...$$

the sum being taken over all integer subsets $(a_1, a_2, a_3, ..., a_l, ...)$ consistent with,

$$\sum_l a_l = N$$

then we shall show the result is also true for *N+1*.

$$(x_1 + x_2 + x_3 + ... + x_l + ...)^{N+1}$$

$$= (x_1 + x_2 + x_3 + ... + x_l + ...) \sum_{a_l} \frac{N!}{a_1! a_2! a_3! .... a_l!...} x_1^{a_1} x_2^{a_2} x_3^{a_3} .... x_l^{a_l} ... =$$

$$x_1 \sum_{a_l} \frac{N!}{a_1! a_2! a_3! .... a_l!...} x_1^{a_1} x_2^{a_2} x_3^{a_3} .... x_l^{a_l} ... + x_2 \sum_{a_l} \frac{N!}{a_1! a_2! a_3! .... a_l!...} x_1^{a_1} x_2^{a_2} x_3^{a_3} .... x_l^{a_l} ... +$$

$$x_3 \sum_{a_l} \frac{N!}{a_1! a_2! a_3! .... a_l!...} x_1^{a_1} x_2^{a_2} x_3^{a_3} .... x_l^{a_l} ... + x_l \sum_{a_l} \frac{N!}{a_1! a_2! a_3! .... a_l!...} x_1^{a_1} x_2^{a_2} x_3^{a_3} .... x_l^{a_l} ... \quad (22)$$

First term on the right hand side of (22) $= x_1 \sum_{a_l} \frac{N!}{a_1! a_2! a_3! .... a_l!...} x_1^{a_1} x_2^{a_2} x_3^{a_3} .... x_l^{a_l} ...$

$$= \sum_{a_l} \frac{N!}{a_1! a_2! a_3! .... a_l!...} x_1^{a_1+1} x_2^{a_2} x_3^{a_3} .... x_l^{a_l} ...$$

$$= \sum_{b_l} \frac{N!}{(b_1 - 1)! b_2! b_3! .... b_l!...} x_1^{b_1} x_2^{b_2} x_3^{b_3} .... x_l^{b_l} ... = \sum_{b_l} \frac{N! b_1}{b_1! b_2! b_3! .... b_l!...} x_1^{b_1} x_2^{b_2} x_3^{b_3} .... x_l^{b_l} ...$$

(22a)

where,
$$b_1 = a_1 + 1, \quad b_2 = a_2, \quad b_3 = a_3, \quad ..., \quad b_l = a_l, ...,$$
and,

$$\sum_l b_l = 1 + \sum_l a_l = 1 + N \quad (22b)$$

second term on the right hand side of (22) $= x_2 \sum_{a_l} \frac{N!}{a_1! a_2! a_3! .... a_l!...} x_1^{a_1} x_2^{a_2} x_3^{a_3} .... x_l^{a_l} ...$

$$= \sum_{a_l} \frac{N!}{a_1! a_2! a_3! .... a_l!...} x_1^{a_1} x_2^{a_2+1} x_3^{a_3} .... x_l^{a_l} ...$$

$$= \sum_{b_l} \frac{N!}{b_1! (b_2 - 1)! b_3! .... b_l!...} x_1^{b_1} x_2^{b_2} x_3^{b_3} .... x_l^{b_l} ... = \sum_{b_l} \frac{N! b_2}{b_1! b_2! b_3! .... b_l!...} x_1^{b_1} x_2^{b_2} x_3^{b_3} .... x_l^{b_l} ...$$

(22c)

where,
$$b_1 = a_1 \quad b_2 = a_2 + 1, \quad b_3 = a_3, \quad ..., \quad b_l = a_l, ...,$$
and,

$$\sum_l b_l = 1 + \sum_l a_l = 1 + N \quad (22d)$$

third term on the right hand side of (22) $= x_3 \sum_{a_l} \dfrac{N!}{a_1! a_2! a_3! \ldots a_l! \ldots} x_1^{a_1} x_2^{a_2} x_3^{a_3} \ldots x_l^{a_l} \ldots$

$$= \sum_{a_l} \dfrac{N!}{a_1! a_2! a_3! \ldots a_l! \ldots} x_1^{a_1} x_2^{a_2} x_3^{a_3+1} \ldots x_l^{a_l} \ldots =$$

$$= \sum_{b_l} \dfrac{N!}{b_1! b_2! (b_3-1)! \ldots b_l! \ldots} x_1^{b_1} x_2^{b_2} x_3^{b_3} \ldots x_l^{b_l} \ldots$$

$$= \sum_{b_l} \dfrac{N! b_3}{b_1! b_2! b_3! \ldots b_l! \ldots} x_1^{b_1} x_2^{b_2} x_3^{b_3} \ldots x_l^{b_l} \ldots \qquad (22e)$$

where,
$$b_1 = a_1 \quad b_2 = a_2 \quad b_3 = a_3 + 1, \quad \ldots, \quad b_l = a_l, \ldots,$$
and,

$$\sum_l b_l = 1 + \sum_l a_l = 1 + N \qquad (22f)$$

l-th term on the right hand side of (22) $= x_l \sum_{a_l} \dfrac{N!}{a_1! a_2! a_3! \ldots a_l! \ldots} x_1^{a_1} x_2^{a_2} x_3^{a_3} \ldots x_l^{a_l} \ldots$

$$= \sum_{a_l} \dfrac{N!}{a_1! a_2! a_3! \ldots a_l! \ldots} x_1^{a_1} x_2^{a_2} x_3^{a_3} \ldots x_l^{a_l+1} \ldots$$

$$= \sum_{b_l} \dfrac{N!}{b_1! b_2! b_3! \ldots (b_l-1)! \ldots} x_1^{b_1} x_2^{b_2} x_3^{b_3} \ldots x_l^{b_l} \ldots$$

$$= \sum_{b_l} \dfrac{N! b_l}{b_1! b_2! b_3! \ldots b_l! \ldots} x_1^{b_1} x_2^{b_2} x_3^{b_3} \ldots x_l^{b_l} \ldots \qquad (22g)$$

where,
$$b_1 = a_1 \quad b_2 = a_2 \quad b_3 = a_3 \quad \ldots, \quad b_l = a_l + 1, \ldots,$$
and,

$$\sum_l b_l = 1 + \sum_l a_l = 1 + N \qquad (22h)$$

From (22a, 22c, 22e, 22g) in conjunction with (22) we have,

$$(x_1 + x_2 + x_3 + \ldots + x_l + \ldots)^{N+1} = \sum_{b_l} \dfrac{N!(b_1 + b_2 + b_3 + \ldots + b_l + \ldots)}{b_1! b_2! b_3! \ldots b_l! \ldots} x_1^{b_1} x_2^{b_2} x_3^{b_3} \ldots x_l^{b_l} \ldots \qquad (22i)$$

where from (22b, 22d, 22f, 22h),

$$(x_1 + x_2 + x_3 + ... + x_l + ...)^{N+1} = \sum_{b_l} \frac{(N+1)!}{b_1! b_2! b_3! .... b_l! ...} x_1^{b_1} x_2^{b_2} x_3^{b_3} .... x_l^{b_l} ... \quad (22j)$$

since $(a_l),(b_l)$ are dummy indices we see that if the multinomial theorem is true for some positive integer N, then from (22j) it is also true for $N+1$. Trivially, the result is also valid for $N=1$. Therefore, by the principle of induction the result is true for all positive integers N. So that for a finite number of terms inside the brackets on the left hand side

$$(x_1 + x_2 + x_3 + ... + x_l + ... + x_n)^N = \sum_{a_l} \frac{N!}{a_1! a_2! a_3! .... a_l! ... a_n!} x_1^{a_1} x_2^{a_2} x_3^{a_3} .... x_l^{a_l} ... x_n^{a_n} \quad (22k)$$

the sum being taken over all integer subsets $(a_1, a_2, a_3, ..., a_l, ... a_n)$ consistent with,

$$\sum_{l=1}^{n} a_l = N \quad (22l)$$

As an additional sidebar we can show that if (22k, 22l) are true for some positive integer n, then we can again use the principle of mathematical induction to prove

$$(x_1 + x_2 + x_3 + ... + x_l + ... + x_n + x_{n+1})^N$$

$$= \sum_{a_l} \frac{N!}{a_1! a_2! a_3! .... a_l! ... a_n! a_{n+1}!} x_1^{a_1} x_2^{a_2} x_3^{a_3} .... x_l^{a_l} ... x_n^{a_n} x_{n+1}^{a_{n+1}} \quad (22m)$$

where,

$$\sum_{l=1}^{n+1} a_l = N \quad (22n)$$

Proof: If the result (22m) is true for some positive integer n, then by the binomial theorem for a positive integral index,

$$(x_1 + x_2 + x_3 + ... + x_l + ... + x_n + x_{n+1})^N = [(x_1 + x_2 + x_3 + ... + x_l + ... + x_n) + x_{n+1}]^N$$

$$= \sum_{r=0}^{N} {}^N C_r (x_1 + x_2 + x_3 + ... + x_l + ... + x_n)^{N-r} x_{n+1}^r$$

$$= \sum_{r=0}^{N} {}^{N}C_r \left[\sum_{a_l} \frac{(N-r)!}{a_1! a_2! a_3! ... a_l! ... a_n!} x_1^{a_1} x_2^{a_2} x_3^{a_3} .... x_l^{a_l} .... x_n^{a_n} \right] x_{n+1}^{r} \qquad (22\text{o})$$

where,

$$\sum_{l=1}^{n} a_l = N - r \qquad (22\text{p})$$

But, $\quad {}^{N}C_r = \dfrac{N!}{r!(N-r)!} \qquad (22\text{q})$

substituting (22q) in (22o) and simplifying,

$$(x_1 + x_2 + x_3 + ... + x_l + ... + x_n + x_{n+1})^N =$$
$$\sum_{r=0}^{N} \sum_{a_l} \frac{N!}{a_1! a_2! a_3! ... a_l! ... a_n! r!} x_1^{a_1} x_2^{a_2} x_3^{a_3} ... x_l^{a_l} ... x_n^{a_n} x_{n+1}^{r} \qquad (22\text{r})$$

Put, $a_{n+1} = r \qquad (22\text{s})$

Te from (22r, 22p)

$$(x_1 + x_2 + x_3 + ... + x_l + ... + x_n + x_{n+1})^N =$$
$$\sum_{r=0}^{N} \sum_{a_l} \frac{N!}{a_1! a_2! a_3! ... a_l! ... a_n! a_{n+1}!} x_1^{a_1} x_2^{a_2} x_3^{a_3} ... x_l^{a_l} ... x_n^{a_n} x_{n+1}^{a_{n+1}} \qquad (22\text{t})$$

with,

$$\sum_{l=1}^{n+1} a_l = N \qquad (22\text{u})$$

Therefore if the results (22k, 22l) are valid for *n* terms within the round brackets, then (22t, 22u) show that the results are also valid for *(n+1)* terms within the brackets. But The result is obviously true for 2 terms x_1, x_2 within the brackets, by the binomial theorem. Then result is therefore proved.

References :

1. J. McConnell – Quantum Particle Dynamics; North-Holland Publishing Company , 1960 , Chapters 7, 8
2. E. Schrödinger – Statistical Thermodynamics; Cambridge University Press, 1962, Chapters 1 – 3.
3. W. Pauli – Pauli Lectures on Physics, Volume 4, Statistical Mechanics, MIT Press, 1973, Chapter 4.
4. F. Bloch – Fundamentals of Statistical Mechanics, Stanford University Press, 1989, Chapters 4, 5.

Q and A Section for Ch 4

1. Show that the necessary and sufficient condition for 2 linear self-adjoint operators to commute is that they have in common a complete set of eigenfunctions.

2. Show that the square of the eigenvalue of any component of the angular momentum operator $\vec{M}$, e.g. M_z does not exceed the eigenvalue of M^2 for a common eigenfunction.
   (a) Hence or otherwise derive the complete set of eigenvalues for M_z and M.
   (b) Write the rules for the addition of 2 independent angular momenta: $\vec{M}_1, \vec{M}_2$

3. Write Dirac's equation for the electron in polar coordinates and hence obtain the fine structure formula for the spectral lines of the hydrogen atom.

4. Derive Liouville's Theorem in statistical mechanics that the volume in phase space is an invariant under a canonical transformation.

5. Derive the Hamiltonian for the electromagnetic field in free space (in the absence of charge or current).
   Hence, write down the total Hamiltonian for an electromagnetic field in the presence of charges.

6. Show $\int_{-\infty}^{\infty} \frac{\sin^2 x}{x^2} dx = \pi$

7. Derive the first, second and third order compound matrix elements for virtual transitions in perturbation theory. Hence calculate the transition probability per unit time.

8. Discuss Dirac's Large Number Hypothesis.

9. Derive the general relativistic cosmological differential equations for (a) particle model, (b) continuum model (Friedmann universes).

10. Derive the Friedmann universes in the (a) absence of a cosmological force (b) Einstein-de Sitter

Solutions :

1. Suppose α, β are 2 linear operators with matrix representations A, B respectively with respect to a given set of basis vectors $\vec{e}_1, \vec{e}_2, ..., \vec{e}_n$ in a vector space, V_n with

$$\alpha \vec{e}_i = \sum_{j=1}^{n} a_{ij} \vec{e}_j \quad ; \quad \beta \vec{e}_i = \sum_{j=1}^{n} b_{ij} \vec{e}_j \quad ; \quad i = 1,2,...,n \quad (1)$$

if in addition, A, B are self adjoint,

$$A = A^\perp ; B = B^\perp \tag{2}$$

and have in common a complete set of eigenfunctions $\psi_1, \psi_2, ..., \psi_n$ then A, B commute.

Proof: Let,
$$A\psi_m = \lambda_m \psi_m \;\; ; \;\; B\psi_m = \mu_m \psi_m$$
so that,
$$(AB - BA)\psi_m = (\lambda_m \mu_m - \mu_m \lambda_m) = 0$$

and with respect to the eigenfunctions as a basis in V_n, the matrix elements of $AB - BA$ are,

$$(AB - BA)_{nm} = \int \psi_n^* (AB - BA) \psi_m \, dq = 0$$

All the elements of the matrix $AB - BA$ are zero and therefore,

$$AB - BA = 0 \qquad\qquad\qquad \textbf{QED}$$

The converse is also true : If 2 self-adjoint operators commute, they have in common a complete set of eigenfunctions.

Proof: For purposes of simplicity, we shall assume all the eigenvalues are distinct.

Let, $(\lambda_1, \lambda_2, ..., \lambda_n)$, $(\mu_1, \mu_2, ..., \mu_n)$ be the eigenvalues of A, B respectively, with eigenfunctions $(\phi_1, \phi_2, ..., \phi_n)$, $(\psi_1, \psi_2, ..., \psi_n)$. Take $(\psi_1, \psi_2, ..., \psi_n)$, the eigenfunctions of B as a basis of V_n. Then any eigenfunction of A can be written as,

$$\phi_k = \sum_{s=1}^{n} c_{ks} \psi_s \tag{3}$$

In addition, we can always take the set of eigenfunctions $(\psi_1, \psi_2, ..., \psi_n)$, as constituting an orthonormal set in V_n.

Then, since by hypothesis A commutes with B, it can be easily shown by induction, that A commutes with any arbitrary polynomial $f(B)$.

$$Af(B) - f(B)A = 0 \tag{4}$$

$$\therefore \quad [Af(B)-f(B)A]\phi_k = 0, \quad \Rightarrow [Af(B)-\lambda_k f(B)]\phi_k = 0$$

Therefore,

$$\sum_{s=1}^{n} c_{ks}[Af(B)-\lambda_k f(B)]\psi_s = 0 \tag{5}$$

$$\therefore \quad \sum_{s=1}^{n} c_{ks}f(\mu_s)[A-\lambda_k I]\psi_s = 0, \text{ where } I = (n \times n) \text{ unit matrix} \tag{6}$$

Since, $\phi_k \neq 0$, $\exists$ some $s = s_0$ such that $c_{ks} \neq 0$. Furthermore, since the polynomial $f(\mu_s)$ is arbitrary, we can choose,

$$f(\mu_s) = 0 \quad \text{if } s \neq s_0$$
$$\neq 0 \quad \text{if } s = s_0$$

Therefore, from (6),

$$c_{ks_0} f(\mu_{s_0})[A-\lambda_k I]\psi_{s_0} = 0$$

This implies,

$$(A-\lambda_k I)\psi_{s_0} = 0 \quad \Rightarrow \quad A\psi_{s_0} = \lambda_k \psi_{s_0}$$

i.e. ψ_{s_0} is also an eigenfunction of A.

In the case of repeated eigenvalues, i.e. degeneracies the same argument can be extended with slight modifications (SP).

2. Consider the commutator,

$$[M^2, M_z] = [M_x^2 + M_y^2 + M_z^2, M_z] = [M_x^2 + M_y^2, M_z] = [M_x^2, M_z] + [M_y^2, M_z]$$

$$= \{M_x[M_x, M_z] + [M_x, M_z]M_x\} + \{M_y[M_y, M_z] + [M_y, M_z]M_y\}$$
$$= \{-i\hbar M_x M_y - M_y M_x\} + \{i\hbar M_y M_x + i\hbar M_x M_y\} = 0 \tag{1}$$

using the definition,

$$\vec{M} = \vec{r} \wedge \vec{p} = -i\hbar \vec{r} \wedge \vec{\nabla} \tag{2}$$

and,

$$[M_y, M_z] = i\hbar M_x \; ; \; [M_z, M_x] = i\hbar M_y \; ; \; [M_x, M_y] = i\hbar M_z \tag{3}$$

This is equivalent to, $\quad \vec{M} \wedge \vec{M} = i\hbar \vec{M} \tag{3a}$

Then from exercise (1) we see that the operators M^2, M_z have a common eigenfunction f, with eigenvalues M'^2, M'_z.

$$(M_x^2 + M_y^2 + M_z^2)f = M'^2 f$$
$$M_z^2 f = M_z'^2 f$$

subtracting,

$$(M_x^2 + M_y^2)f = (M'^2 - M_z'^2)f \qquad (4)$$

We now prove that the eigenvalues of $M_x^2 + M_y^2$ are positive.

Proof: Since the operator $M_x^2 + M_y^2$ is self-adjoint, its eigenvalues are real. If ϕ is an eigenvector corresponding to the eigenvalue λ,

Then, $\qquad (M_x^2 + M_y^2)\phi = \lambda \phi$

$$\phi^\perp (M_x^2 + M_y^2)\phi = \lambda \phi^\perp \phi$$

$\therefore \quad \phi^\perp \{(M_x + iM_y)(M_x - iM_y) + (M_x - iM_y)(M_x + iM_y)\}\phi = 2\lambda \phi^\perp \phi$

Therefore,
$$\{(M_x - iM_y)\phi\}^\perp \{(M_x - iM_y)\phi\} + \{(M_x + iM_y)\phi\}^\perp \{(M_x + iM_y)\phi\} = 2\lambda \phi^\perp \phi \quad (5)$$

In Hilbert space if ϕ is an infinite dimensional column vector with components $\phi_1, \phi_2, ..., \phi_n, ...$ Then $\phi^\perp$ is an infinite dimensional row vector with components $\phi^*_1, \phi^*_2, ..., \phi^*_n, ...$, where a (*) denotes complex conjugation.

$$\therefore \phi^\perp \phi = \sum_{n=1}^{\infty} |\phi_n|^2 > 0$$

Similarly,
$$\{(M_x - iM_y)\phi\}^\perp \{(M_x - iM_y)\phi\} > 0$$

$$\{(M_x + iM_y)\phi\}^\perp \{(M_x + iM_y)\phi\} > 0$$

Therefore (5) implies, $\lambda \geq 0$ **QED**

From (4),

$$|M_z'| \leq M' \qquad (6)$$

(a) We state that the Hamiltonian H, M^2, M_z form a complete set of operators, meaning that the system is fully described by the eigenvalues of the members of the set. More generally, we say that the eigenfunctions f apply not only to M^2, M_z but also to other independent commuting operators which can be written generically as

α, with corresponding eigenvalues α'.
Again,
$$M^2(M_x + iM_y)f = (M_x + iM_y)M^2 f = (M_x + iM_y)M'^2 f = M'^2(M_x + iM_y)f$$
and,
$$\alpha(M_x + iM_y)f = (M_x + iM_y)\alpha f = (M_x + iM_y)\alpha' f = \alpha'(M_x + iM_y)f$$

So that $(M_x + iM_y)f$ is an eigenfunction of both M^2, α with eigenvalues M'^2, α'. But using (3),

$$M_z(M_x + iM_y)f = \{(M_x M_z + i\hbar M_y) + i(M_y M_z - i\hbar M_x)\}f$$
$$= (M'_z + \hbar)(M_x + iM_y)f \tag{7}$$

Therefore, we have for the operator M_z the following sequence of eigenvectors and eigenvalues: These results can be proved using induction.

Eigenvectors: $\quad f, \quad (M_x + iM_y)f, \quad (M_x + iM_y)^2 f, \quad (M_x + iM_y)^3 f, \ldots,$
Eigenvalues: $\quad M'_z, \quad (M'_z + \hbar), \quad (M'_z + 2\hbar), \quad (M'_z + 3\hbar), \ldots,$

Proof: We prove this result by induction. Suppose that for some value of the positive integer n, $(M_x + iM_y)^n f$ is an eigenvector of M_z with eigenvalue $(M'_z + n\hbar)$. Then using (3),
$$M_z(M_x + iM_y)^n f = (M'_z + n\hbar)(M_x + iM_y)^n f$$
$$\therefore M_z(M_x + iM_y)^{n+1} f = M_z(M_x + iM_y)(M_x + iM_y)^n f$$
$$= \{(i\hbar M_y + M_x M_z) + i(M_y M_z - i\hbar M_x)\}(M_x + iM_y)^n f$$
$$= \{(M_x + iM_y)M_z + \hbar(M_x + iM_y)\}(M_x + iM_y)^n f$$
$$= \{(M_x + iM_y)(M'_z + n\hbar)(M_x + iM_y)^n f + \hbar(M_x + iM_y)^{n+1} f\}$$
$$= \{M'_z + (n+1)\hbar\}(M_x + iM_y)^{n+1} f$$

Therefore if the result is true for some value of n, then it is also true for $n+1$. But it is certainly true for $n = 0$. Therefore the result is true for all n. **QED**
The above sequence will terminate when condition (6) is violated.
Therefore,

$\exists$ a positve integer N such that

$$M'_z + N\hbar = M' \tag{8}$$
and,
$$(M_x + iM_y)^N f \neq 0 \ ; \ (M_x + iM_y)^{N+1} f = 0 \tag{9}$$

From (9),
$$(M_x - iM_y)(M_x + iM_y)^{N+1} f = 0$$
i.e.
$$(M_x - iM_y)(M_x + iM_y)(M_x + iM_y)^N f = 0$$
From (3),
$$(M_x^2 + M_y^2 - \hbar M_z)(M_x + iM_y)^N f = 0$$
$$\therefore \quad (M^2 - M_z^2 - \hbar M_z)(M_x + iM_y)^N f = 0$$
$$\therefore \quad \{M'^2 - (M_z' + N\hbar)^2 - \hbar(M_z' + N\hbar)\}(M_x + iM_y)^N f = 0$$
$$\therefore \text{ From (9),} \quad M'^2 - (M_z' + N\hbar)^2 - \hbar(M_z' + N\hbar) = 0 \qquad (10)$$

Similarly, we can easily show that the following sequence of eigenvectors and Eigenvalues of M_z must also terminate in accordance with condition (6).

Eigenvectors: f, $(M_x - iM_y)f$, $(M_x - iM_y)^2 f$, $(M_x - iM_y)^3 f$, ...,
Eigenvalues: M_z', $(M_z' - \hbar)$, $(M_z' - 2\hbar)$, $(M_z' - 3\hbar)$, ...,

We have a result similar to (10), with a corresponding nonnegative integer Λ, satisfying,
$$M'^2 - (M_z' - \Lambda\hbar)^2 + \hbar(M_z' - \Lambda\hbar) = 0 \qquad (11)$$
Subtracting (10, 11),
$$\{(M_z' + N\hbar)^2 - (M_z' - \Lambda\hbar)^2\} + \hbar\{(M_z' + N\hbar) + (M_z' - \Lambda\hbar)\} = 0$$
Simplifying,
$$\hbar\{(M_z' + N\hbar) + (M_z' - \Lambda\hbar)\}\{N + \Lambda + 1\} = 0 \qquad (12)$$

$$\therefore \quad M_z' - \Lambda\hbar = -(M_z' + N\hbar) \qquad (13)$$

$$\therefore (M_z' + N\hbar) - (M_z' - \Lambda\hbar) = (N + \Lambda)\hbar \Rightarrow M_z' + N\hbar = \frac{N+\Lambda}{2}\hbar = j\hbar \qquad (14)$$
where,
$$j = \frac{N+\Lambda}{2} \qquad (15)$$

From (10, 14),
$$M'^2 - j^2\hbar^2 - j\hbar^2 = 0$$

$$\therefore \quad M' = \hbar\sqrt{j(j+1)} \qquad (16)$$

From (15), $j = 0, \frac{1}{2}, 1, \frac{3}{2}, 2, ...$

The observed eigenvalues of M_z are:

$$\{M'_z, (M'_z + \hbar), (M'_z + 2\hbar), (M'_z + 3\hbar), ..., (M'_z + N\hbar)\} \ \& \ \{(M'_z - \hbar), (M'_z - 2\hbar), (M'_z - 3\hbar), ..., (M'_z - \Lambda\hbar)\}$$

There are → $N + 1 + \Lambda = 2j + 1$ eigenvalues of $M_z = \{-j\hbar, ..., j\hbar\}$

(b) Addition of angular momenta :

$$\vec{M}_1 \wedge \vec{M}_1 = i\hbar \vec{M}_1 \ ; \ \vec{M}_2 \wedge \vec{M}_2 = i\hbar \vec{M}_2 \tag{17}$$

$$\vec{M} = M_1 + M_2 \tag{18}$$

$$\therefore \quad \vec{M} \wedge \vec{M} = (\vec{M}_1 + \vec{M}_2) \wedge (\vec{M}_1 + \vec{M}_2)$$

$$= \vec{M}_1 \wedge \vec{M}_1 + \vec{M}_2 \wedge \vec{M}_1 + \vec{M}_1 \wedge \vec{M}_2 + \vec{M}_2 \wedge \vec{M}_2$$

But since $\vec{M}_1, \vec{M}_2$ are independent,

$$\vec{M}_1 \wedge \vec{M}_2 = \vec{M}_2 \wedge \vec{M}_1 = 0 \tag{19}$$

$$\therefore \quad \vec{M} \wedge \vec{M} = i\hbar(\vec{M}_1 + \vec{M}_2) = i\hbar\vec{M} \tag{20}$$

Therefore, the sum of 2 independent angular momenta, $\vec{M}_1, \vec{M}_2$ is also an angular Momentum. The system is fully described by the set of commuting self-adjoint operators $\{M_1^2, M_2^2, M_{1z}, M_{2z}, \beta\}$. The eigenvectors of this set are written generically as,

$$|j_1, j_2, m_1, m_2, \beta'>$$

where from exercise 2(a) we see that this corresponds to the eigenstate in which $M_1^2, M_2^2, M_{1z}, M_{2z}, \beta$ have respectively the eigenvalues:

$\{\hbar^2 j_1(j_1+1), \hbar^2 j_2(j_2+1), \hbar m_1, \hbar m_2, \beta'\}$. For a given j_1, j_2 there will respectively be $(2j_1 + 1)$ values of m_1, and $(2j_2 + 1)$ values of m_2. $|m_1| \leq j_1 \ ; \ |m_2| \leq j_2$

$$\therefore \quad M_z |j_1, j_2, m_1, m_2, \beta'> = (M_{1z} + M_{2z})|j_1, j_2, m_1, m_2, \beta'>$$
$$= \hbar(m_1 + m_2)|j_1, j_2, m_1, m_2, \beta'>$$

$$= \hbar m \mid j_1, j_2, m_1, m_2, \beta' > \qquad (21)$$

where,
$$m = m_1 + m_2$$

with m_1, m_2 corresponding to:

$$m_1 = -j_1, \; -(j_1-1), \; -(j_1-2), \ldots, (j_1-2), \; (j_1-1), \; j_1$$
$$m_2 = -j_2, \; -(j_2-1), \; -(j_2-2), \ldots, (j_2-2), \; (j_2-1), \; j_2 \qquad (22)$$

and the eigenvalues $m\hbar$ and eigenvectors of M_z correspond to:

$$
\begin{aligned}
m &= -(j_1 + j_2) \rightarrow \mid j_1, j_2, -j_1, -j_2, \beta' > \\
&= (j_1 + j_2) \rightarrow \mid j_1, j_2, j_1, j_2, \beta' > \\
\\
&= -(j_1 + j_2 - 1) \rightarrow \mid j_1, j_2, -j_1, -(j_2-1), \beta' >, \mid j_1, j_2, -(j_1-1), -j_2, \beta' > \\
&= (j_1 + j_2 - 1) \rightarrow \mid j_1, j_2, j_1, (j_2-1), \beta' >, \mid j_1, j_2, (j_1-1), j_2, \beta' > \\
\\
&= -(j_1 + j_2 - 2) \rightarrow \mid j_1, j_2, -j_1, -(j_2-2), \beta' >, \mid j_1, j_2, -(j_1-2), -j_2, \beta' >, \\
&\qquad \mid j_1, j_2, -(j_1-1), -(j_2-1), \beta' > \\
&= (j_1 + j_2 - 2) \rightarrow \mid j_1, j_2, j_1, (j_2-2), \beta' >, \mid j_1, j_2, (j_1-2), j_2, \beta' >, \\
&\qquad \mid j_1, j_2, (j_1-1), (j_2-1), \beta' > \\
&= -(j_1 + j_2 - 3) \rightarrow \mid j_1, j_2, -j_1, -(j_2-3), \beta' >, \mid j_1, j_2, -(j_1-3), -j_2, \beta' >, \\
&\qquad \mid j_1, j_2, -(j_1-1), -(j_2-2), \beta' >, \mid j_1, j_2, -(j_1-2), -(j_2-1), \beta' > \\
&= (j_1 + j_2 - 3) \rightarrow \mid j_1, j_2, j_1, (j_2-3), \beta' >, \mid j_1, j_2, (j_1-3), j_2, \beta' >, \\
&\qquad \mid j_1, j_2, (j_1-1), (j_2-2), \beta' >, \mid j_1, j_2, (j_1-2), (j_2-1), \beta' >
\end{aligned}
$$

$$(23)$$

of M are given by,

$$M' = \hbar \sqrt{j(j+1)} \quad \text{where,} \quad j \geq |m| \qquad (24)$$

From (23),

$$j = j_1 + j_2, \; j_1 + j_2 - 1, \; j_1 + j_2 - 2, \; j_1 + j_2 - 3, \qquad (25)$$

This section is to be studied in parallel with the answers to question 6 in the chapter on wave particle duality (Chapter 1)

Define the operator j as follows,

$$j\hbar = \alpha_0 [\vec{\pi} \bullet \vec{m} + \hbar] \qquad (1)$$

where $\vec{m}$ is the angular momentum operator,

$$\vec{m} = \vec{r} \wedge \vec{p} = -i\hbar \; \vec{r} \wedge \vec{\nabla} \tag{2}$$

Then using the solutions (3 & 9) of chapter 1,

$$\therefore \quad j^2\hbar^2 = [\vec{\pi} \bullet \vec{m} + \hbar]^2 = (\vec{\pi} \bullet \vec{m})^2 + 2\hbar(\vec{\pi} \bullet \vec{m}) + \hbar^2 \tag{3}$$

Using solution (14),

$$(\vec{\pi} \bullet \vec{m})^2 = (\vec{\pi} \bullet \vec{m})(\vec{\pi} \bullet \vec{m}) = \vec{m} \bullet \vec{m} + i[\vec{\pi} \bullet (\vec{m} \wedge \vec{m})] \tag{4}$$

But, $\quad \vec{m} \wedge \vec{m} = i\hbar\vec{m} \tag{5}$

Proof: Using suffix notation in conjunction with equation (2) and the permutation tensor,

$$(\vec{m} \wedge \vec{m})_w = \varepsilon_{wqr} m_q m_r = \varepsilon_{wqr}(\varepsilon_{qst} x_s p_t)(\varepsilon_{ruv} x_u p_v) = -\varepsilon_{qwr}\varepsilon_{qst}\varepsilon_{ruv} x_s p_t x_u p_v$$

$$= -\varepsilon_{ruv}(\delta_{ws}\delta_{rt} - \delta_{wt}\delta_{rs}) x_s p_t x_u p_v = -\varepsilon_{ruv}(x_w p_r - x_r p_w) x_u p_v$$

$$= i\hbar\varepsilon_{ruv}(x_w \frac{\partial}{\partial x_r} - x_r \frac{\partial}{\partial x_w}) x_u p_v = i\hbar\varepsilon_{ruv}(x_w \delta_{ru} p_v - x_r \delta_{uw} p_v)$$

$$= i\hbar(\varepsilon_{uuv} x_w p_v - \varepsilon_{rwv} x_r p_v) = i\hbar\varepsilon_{wrv} x_r p_v \quad (\because \varepsilon_{uuv} = 0)$$

$$= i\hbar(\vec{r} \wedge \vec{p})_w \quad\quad\quad\quad\quad \textbf{QED}$$

Therefore, from (3, 4, 5)

$$j^2\hbar^2 = |m|^2 + \hbar(\vec{\pi} \bullet \vec{m}) + \hbar^2$$

$$= |\vec{m} + \frac{1}{2}\hbar\vec{\pi}|^2 + \frac{1}{4}\hbar^2 \quad (\because \pi^2 = \pi_1^2 + \pi_2^2 + \pi_3^2 = 3)$$

$$= M^2 + \frac{1}{4}\hbar^2 \tag{6}$$

From exercise 2, we know the eigenvalues of the z- component of the orbital angular momentum $m_z = m_1\hbar$, where m_1 is an integer, such that $|m_1| \leq j_1$. From the exercises in Chapter 1, we saw that the eigenvalues of π_z are ± 1.

Therefore, the eigenvalues of $M_z = -(j_1 + \frac{1}{2})\hbar, \ldots, (j_1 + \frac{1}{2})\hbar$

The eigenvalues of $M^2 = l(l+1)\hbar^2$ where,

$$l = j_1 + \frac{1}{2}, \quad j_1 - \frac{1}{2}$$

Therefore the eigenvalues of j^2 in (6) are integers.

To transform to spherical polar coordinates, define the operators,

$$p_r = r^{-1}\{(\vec{r} \bullet \vec{p}) - i\hbar\} \tag{7a}$$

$$r\varepsilon = \vec{\alpha} \bullet \vec{r} \tag{7b}$$

From (7a),

$$p_r = r^{-1}(xp_x + yp_y + zp_z - i\hbar) = -i\hbar r^{-1}\left(x\frac{\partial}{\partial x} + y\frac{\partial}{\partial y} + z\frac{\partial}{\partial z} + 1\right) \tag{7c}$$

But,

$$x = r\sin\theta\cos\phi, \quad y = r\sin\theta\sin\phi, \quad z = r\cos\theta$$

$$\therefore \frac{\partial r}{\partial x} = \frac{x}{r}, \quad \frac{\partial r}{\partial y} = \frac{y}{r}, \quad \frac{\partial r}{\partial z} = \frac{z}{r};$$

$$\frac{\partial \theta}{\partial x} = \frac{zx}{r^3 \sin\theta}, \quad \frac{\partial \theta}{\partial y} = \frac{zy}{r^3 \sin\theta}, \quad \frac{\partial \theta}{\partial z} = \frac{-(x^2+y^2)}{r^3 \sin\theta}$$

$$\frac{\partial \phi}{\partial x} = \frac{-y}{r^2 \sin^2\theta}, \quad \frac{\partial \phi}{\partial y} = \frac{x}{r^2 \sin^2\theta}, \quad \frac{\partial \phi}{\partial z} = 0$$

$$\therefore x\frac{\partial}{\partial x} + y\frac{\partial}{\partial y} + z\frac{\partial}{\partial z} =$$

$$x\left(\frac{\partial r}{\partial x}\frac{\partial}{\partial r} + \frac{\partial \theta}{\partial x}\frac{\partial}{\partial \theta} + \frac{\partial \phi}{\partial x}\frac{\partial}{\partial \phi}\right) + y\left(\frac{\partial r}{\partial y}\frac{\partial}{\partial r} + \frac{\partial \theta}{\partial y}\frac{\partial}{\partial \theta} + \frac{\partial \phi}{\partial y}\frac{\partial}{\partial \phi}\right) +$$

$$z\left(\frac{\partial r}{\partial z}\frac{\partial}{\partial r} + \frac{\partial \theta}{\partial z}\frac{\partial}{\partial \theta} + \frac{\partial \phi}{\partial z}\frac{\partial}{\partial \phi}\right) = r\left(\frac{\partial}{\partial r}\right) \tag{7d}$$

$$\therefore p_r = -i\hbar\left(\frac{\partial}{\partial r} + \frac{1}{r}\right) \tag{7e}$$

p_r, ε are self adjoint operators.

Proof: $p_r^\perp = (p_x x + p_y y + p_z z + i\hbar) r^{-1} = (x p_x + y p_y + z p_z + 2i\hbar) r^{-1}$

$$= (r p_r - i\hbar) r^{-1} \quad \text{, from (7c)}$$

$$r[r, p_r] = r(r p_r - p_r r) = r^2 p_r - r p_r r = [r, r p_r]$$

$$= [r, x p_x + y p_y + z p_z - i\hbar] = [r, x p_x] + [r, y p_y] + [r, z p_z] \quad (7f)$$

But, $[r, x p_x] = r x p_x - x p_x r = x r p_x - x p_x r = x(r p_x - p_x r)$

$$p_x r = -i\hbar \frac{\partial}{\partial x} r = -i\hbar \left(r \frac{\partial}{\partial x} + \frac{\partial r}{\partial x} \right) = r p_x - i\hbar \frac{x}{r}$$

$$\therefore \quad [r, x p_x] = i\hbar \frac{x^2}{r}$$

Similarly,

$$[r, y p_y] = i\hbar \frac{y^2}{r} \quad , \quad [r, z p_z] = i\hbar \frac{z^2}{r}$$

Therefore (7f) reduces to,

$$r[r, p_r] = i\hbar r \tag{7g}$$

or, $\quad [r, p_r] = i\hbar \quad \Rightarrow \quad r p_r - p_r r = i\hbar \tag{7h}$

$$\therefore \quad p_r^\perp = p_r r r^{-1} = p_r \qquad \textbf{QED}$$

Proof: From (7b),

$$\varepsilon r = r^{-1} (\vec{\alpha} \bullet \vec{r}) r = r^{-1} r (\vec{\alpha} \bullet \vec{r}) = \vec{\alpha} \bullet \vec{r} = r\varepsilon \tag{7h'}$$

$$\varepsilon^\perp r = (r\varepsilon)^\perp = (\vec{\alpha} \bullet \vec{r})^\perp = (\vec{\alpha} \bullet \vec{r}) = r\varepsilon = \varepsilon r$$

$$\varepsilon^\perp = \varepsilon \qquad \textbf{QED}$$

We now prove a number of necessary propositions: The orbital angular momentum is defined by,

$$m = \vec{r} \wedge \vec{p} = \begin{pmatrix} x, y, z \\ p_x, p_y, p_z \end{pmatrix} = (yp_z - zp_y, zp_x - xp_z, xp_y - yp_x) \quad (7i)$$

Then the commutators,

$$[x, m_x] = [x, yp_z - zp_y] = [x, yp_z] - [x, zp_y] = y[x, p_z] - z[x, p_y] = 0 \quad (7j)$$

$$[y, m_x] = [y, yp_z - zp_y] = [y, yp_z] - [y, zp_y] = -z[y, p_y] = -i\hbar z \quad (7k)$$

$$[z, m_x] = [z, yp_z - zp_y] = [z, yp_z] - [z, zp_y] = y[z, p_z] = i\hbar y \quad (7l)$$

$$[r, m_x] = [r, yp_z - zp_y] = [r, yp_z] - [r, zp_y] = -i\hbar [r, y\frac{\partial}{\partial z}] + i\hbar [r, z\frac{\partial}{\partial y}]$$

$$= i\hbar y \frac{\partial r}{\partial z} - i\hbar z \frac{\partial r}{\partial y} = 0 \quad (7m)$$

By induction we can easily show that for any arbitrary function of r, $f(r)$
$$[f(r), m_x] = 0 \quad (7n)$$

Again,
$$[1, m_x] = [rr^{-1}, m_x] = 0$$
$$\therefore rr^{-1}m_x - m_x rr^{-1} = 0 \qquad (yp_z - zp_y, zp_x - xp_z, xp_y - yp_x)$$
$$r(r^{-1}m_x - m_x r^{-1}) + (rm_x r^{-1} - m_x rr^{-1}) = 0$$

i.e. $r[r^{-1}, m_x] + [r, m_x]r^{-1} = 0 \Rightarrow r[r^{-1}, m_x] = 0$, from (7m) \quad (7o)

Therefore $\vec{m}$ commutes with r^{-1}. Consider the following set of commutators,

$$[p_x, m_x] = [p_x, yp_z - zp_y] = [p_x, yp_z] - [p_x, zp_y] = y[p_x, p_z] - z[p_x, p_y] = 0 \quad (7p)$$
$$[p_y, m_x] = [p_y, yp_z - zp_y] = [p_y, yp_z] - [p_y, zp_y] = [p_y, y]p_z = -i\hbar p_z \quad (7q)$$
$$[p_z, m_x] = [p_z, yp_z - zp_y] = [p_z, yp_z] - [p_z, zp_y] = -[p_z, z]p_y = i\hbar p_y \quad (7r)$$

From (7a), (7j-7l, 7p-7r),

$$[rp_r, m_x] = [xp_x + yp_y + zp_z - i\hbar, m_x] = [xp_x, m_x] + [yp_y, m_x] + [zp_z, m_z]$$
$$= (x[p_x, m_x] + [x, m_x]p_x) + (y[p_y, m_x] + [y, m_x]p_y) + (z[p_z, m_x] + [z, m_x]p_z)$$
$$= -i\hbar yp_z - i\hbar zp_y + i\hbar zp_y + i\hbar yp_z = 0$$

$$\therefore \quad [rp_r, m_x] = 0$$

i.e. $r[p_r, m_x] + [r, m_x]p_r = 0$

$\therefore \quad r[p_r, m_x] = 0$, from (7m)

$\therefore \quad [p_r, \vec{m}] = 0$ \hfill (7s)

Lemma 1 : p_r commutes with α_0

Proof : $p_r \alpha_0 = r^{-1}\{x_l p_l - i\hbar\}\alpha_0 = r^{-1}\{x_l p_l \alpha_0 - i\hbar \alpha_0\}$

But α_0 is independent of the coordinates, time, momentum and energy.

$$p_r \alpha_0 = \alpha_0 r^{-1}\{x_l p_l - i\hbar\} = \alpha_0 p_r \qquad \textbf{QED} \qquad (7t)$$

Lemma 2 : p_r commutes with $\vec{\pi}$. Since $\vec{\pi}$ depends only on $\vec{\alpha}$, from equation (9) of exercise 6(a) of chapter 1, the result follows at once, since $\vec{\alpha}$, by hypothesis commutes with the coordinate and momentum operators $\vec{r}, \vec{p}$. Therefore from (1), (7s), p_r commutes with j.

ε commutes with j.
Proof: From (1, 7b),
$$(\varepsilon j - j\varepsilon)\hbar = r^{-1}(\vec{\alpha} \bullet \vec{r})\alpha_0\{\vec{\pi} \bullet \vec{m} + \hbar\} - \alpha_0\{\vec{\pi} \bullet \vec{m} + \hbar\}r^{-1}(\vec{\alpha} \bullet \vec{r})$$
$$= 2\hbar r^{-1}(\vec{\alpha} \bullet \vec{r})\alpha_0 - r^{-1}\alpha_0\{(\vec{\pi} \bullet \vec{m})(\vec{\alpha} \bullet \vec{r}) + (\vec{\alpha} \bullet \vec{r})(\vec{\pi} \bullet \vec{m})\} \qquad (7u)$$

But, from equations (16) of exercise 6(a) of chapter 1, with a slight change of notation,

$$\alpha_k = \rho \pi_k \; ; k=1, 2, 3 \qquad (7v)$$

Pre-multiplying by ρ,

$$\pi_k = \rho \alpha_k, \qquad \qquad \because \rho^2 = 1 \qquad (7w)$$

This can be re-written as, $\vec{\pi} = \rho \vec{\alpha}$ \hfill (7x)

$\therefore \quad (\vec{\alpha} \bullet \vec{r})(\vec{\pi} \bullet \vec{m}) = \rho(\vec{\alpha} \bullet \vec{r})(\vec{\alpha} \bullet \vec{m})$ \hfill (7y)

because ρ commutes with $\vec{\alpha}$. For example, using the anticommuting features of the α operators

$$\rho \alpha_1 = -i\alpha_1 \alpha_2 \alpha_3 \alpha_1 = i\alpha_1 \alpha_2 \alpha_1 \alpha_3 = -i\alpha_1 \alpha_1 \alpha_2 \alpha_3 = -i\alpha_2 \alpha_3$$
$$\alpha_1 \rho = -i\alpha_1 \alpha_1 \alpha_2 \alpha_3 = -i\alpha_2 \alpha_3$$

$$\therefore \quad \rho\alpha_1 = \alpha_1 \rho$$

Similarly, we can show : $\rho\alpha_2 = \alpha_2 \rho$; $\rho\alpha_3 = \alpha_3 \rho$

Again, from equations (14) of exercise 6(a) of chapter 1, and (7x)

$$(\vec{\pi} \bullet \vec{a})(\vec{\pi} \bullet \vec{b}) = \vec{a} \bullet \vec{b} + i\vec{\pi} \bullet (\vec{a} \wedge \vec{b})$$

$$(\rho\vec{\alpha} \bullet \vec{a})(\rho\vec{\alpha} \bullet \vec{b}) = \vec{a} \bullet \vec{b} + i\vec{\pi} \bullet (\vec{a} \wedge \vec{b})$$

This implies,
$$(\vec{\alpha} \bullet \vec{a})(\vec{\alpha} \bullet \vec{b}) = \vec{a} \bullet \vec{b} + i\vec{\pi} \bullet (\vec{a} \wedge \vec{b}) \tag{7z}$$

From (7y, 7z),

$$(\vec{\alpha} \bullet \vec{r})(\vec{\pi} \bullet \vec{m}) = \rho\{(\vec{r} \bullet \vec{m}) + i\vec{\pi} \bullet (\vec{r} \wedge \vec{m})\} \tag{8}$$

Again, $(\vec{\pi} \bullet \vec{m})(\vec{\alpha} \bullet \vec{r}) = \rho(\vec{\alpha} \bullet \vec{m})(\vec{\alpha} \bullet \vec{r}) = \rho\{(\vec{m} \bullet \vec{r}) + i\vec{\pi} \bullet (\vec{m} \wedge \vec{r})\}$ (9)

$$\vec{r} \bullet \vec{m} = x(yp_z - zp_y) + y(zp_x - xp_z) + z(xp_y - yp_x) = 0$$
$$\vec{m} \bullet \vec{r} = (yp_z - zp_y)x + (zp_x - xp_z)y + (xp_y - yp_x)z = 0$$

Therefore, from (7x, 8, 9)

$$(\vec{\pi} \bullet \vec{m})(\vec{\alpha} \bullet \vec{r}) + (\vec{\alpha} \bullet \vec{r})(\vec{\pi} \bullet \vec{m}) = \rho\{i\vec{\pi} \bullet (\vec{m} \wedge \vec{r}) + i\vec{\pi} \bullet (\vec{r} \wedge \vec{m})\}$$
$$= i\vec{\alpha} \bullet (\vec{m} \wedge \vec{r} + \vec{r} \wedge \vec{m}) \tag{10}$$

Using suffix notation,

$$(\vec{m} \wedge \vec{r} + \vec{r} \wedge \vec{m})_i = \varepsilon_{ijk}(m_j x_k + x_j m_k) = \varepsilon_{ijk} m_j x_k + \varepsilon_{ijk} x_j m_k$$
$$= \varepsilon_{ijk} m_j x_k + \varepsilon_{ikj} x_k m_j$$
$$= \varepsilon_{ijk} m_j x_k - \varepsilon_{ijk} x_k m_j$$
$$= \varepsilon_{ijk}[m_j, x_k]$$

when $i = 1$, using (7j-7l) in conjunction with similar results

$[x, m_y] = i\hbar z$	(7j')
$[y, m_y] = 0$	(7k')
$[z, m_y] = -i\hbar x$	(7l')
$[x, m_z] = -i\hbar y$	(7j")
$[y, m_z] = i\hbar x$	(7k")
$[z, m_z] = 0$	(7l")

we have,
$$(\vec{m}\wedge\vec{r}+\vec{r}\wedge\vec{m})_1 = [m_2, x_3]-[m_3, x_2] = -[z, m_y]+[y, m_z]$$
$$= 2i\hbar x \qquad (10a)$$

Similarly,
$$(\vec{m}\wedge\vec{r}+\vec{r}\wedge\vec{m})_2 = 2i\hbar y \;\; ; \; (\vec{m}\wedge\vec{r}+\vec{r}\wedge\vec{m})_3 = 2i\hbar z \qquad (10b)$$

From (7u, 10, 10a, 10b)
$$\therefore \; (\varepsilon j - j\varepsilon)\hbar = 2\hbar r^{-1}(\vec{\alpha}\bullet\vec{r})\alpha_0 + 2\hbar r^{-1}\alpha_0(\vec{\alpha}\bullet\vec{r}) = 0 \qquad \textbf{QED}$$

Show that ε and p_r commute.

Proof:
$$x(\vec{r}\bullet\vec{p}) - (\vec{r}\bullet\vec{p})x = x(xp_x + yp_y + zp_z) - (xp_x + yp_y + zp_z)x$$
$$= x(xp_x - p_x x) = i\hbar x$$

$$\therefore \; \vec{r}(\vec{r}\bullet\vec{p}) - (\vec{r}\bullet\vec{p})\vec{r} = i\hbar\vec{r}$$

$$\therefore \; i\hbar\vec{\alpha}\bullet\vec{r} = \vec{\alpha}\bullet\{\vec{r}(\vec{r}\bullet\vec{p}) - (\vec{r}\bullet\vec{p})\vec{r}\} = (\vec{\alpha}\bullet\vec{r})(\vec{r}\bullet\vec{p}) - (\vec{r}\bullet\vec{p})(\vec{\alpha}\bullet\vec{r}) \quad (11)$$

From (7a, 7b, 7h),
$$i\hbar r\varepsilon = i\hbar(\vec{\alpha}\bullet\vec{r}) = (\vec{\alpha}\bullet\vec{r})(rp_r + i\hbar) - (rp_r + i\hbar)(\vec{\alpha}\bullet\vec{r})$$
$$= r\varepsilon(p_r r + 2i\hbar) - (rp_r + i\hbar)r\varepsilon$$
$$= r\varepsilon p_r r - rp_r r\varepsilon + i\hbar r\varepsilon$$

$$\therefore \qquad r\varepsilon p_r r - rp_r r\varepsilon = 0$$

But from (7h'), $\quad r(\varepsilon p_r - p_r \varepsilon)r = 0$

$$\therefore \qquad (\varepsilon p_r - p_r \varepsilon) = 0 \qquad \textbf{QED}$$

From (7a , 7b),
$$\varepsilon p_r + i\hbar r^{-1}\varepsilon\alpha_0 j = r^{-1}(\vec{\alpha}\bullet\vec{r})r^{-1}\{(\vec{r}\bullet\vec{p}) - i\hbar\} + ir^{-1}r^{-1}(\vec{\alpha}\bullet\vec{r})\alpha_0^2\{\vec{\pi}\bullet\vec{m}+\hbar\}$$
$$= r^{-2}(\vec{\alpha}\bullet\vec{r})\{\vec{r}\bullet\vec{p} + i(\vec{\pi}\bullet\vec{m})\}$$
$$= r^{-2}(\vec{\alpha}\bullet\vec{r})\{\vec{r}\bullet\vec{p} + i\vec{\pi}\bullet(\vec{r}\wedge\vec{p})\}$$
$$= r^{-2}(\vec{\alpha}\bullet\vec{r})(\vec{\alpha}\bullet\vec{r})(\vec{\alpha}\bullet\vec{p}) \;\;, \;\; \text{from (7z)}$$
$$= \vec{\alpha}\bullet\vec{p} \qquad (12)$$

because, from (7z)
$$(\vec{\alpha}\bullet\vec{r})(\vec{\alpha}\bullet\vec{r}) = \vec{r}\bullet\vec{r} + i\vec{\pi}\bullet(\vec{r}\wedge\vec{r}) = r^2$$

Therefore, the Dirac Hamiltonian for a particle of charge $-e$ is given by,

$$H = c\left[\vec{\alpha}\bullet(\vec{p}+\frac{e}{c}\vec{A})\right]+\alpha_0 m_0 c^2 - eV(r) \tag{13}$$

From (12),

$$H = c\varepsilon p_r + ci\hbar r^{-1}\varepsilon\alpha_0 j + e\vec{\alpha}\bullet\vec{A}+\alpha_0 m_0 c^2 - eV(r) \tag{14}$$

Suppose there is a magnetic field $\vec{H}$ in the z-direction.

$$\vec{H} = \vec{\nabla}\wedge\vec{A} = \left(\frac{\partial A_z}{\partial y}-\frac{\partial A_y}{\partial z},\frac{\partial A_x}{\partial z}-\frac{\partial A_z}{\partial x},\frac{\partial A_y}{\partial x}-\frac{\partial A_x}{\partial y}\right) \tag{15}$$

This is satisfied by,

$$A_x = -\frac{1}{2}Hy \;,\; A_y = \frac{1}{2}Hx \;,\; A_z = 0 \tag{16}$$

Therefore from (14),

$$H = c\varepsilon p_r + ci\hbar r^{-1}\varepsilon\alpha_0 j + \frac{eH}{2}(-y\alpha_x + x\alpha_y)+\alpha_0 m_0 c^2 - eV(r) \tag{17}$$

Again, from (7b, 7h'),

$$\varepsilon^2 = r^{-2}r^2\varepsilon^2 = r^{-2}r\varepsilon r\varepsilon = r^{-2}(r\varepsilon)^2 = r^{-2}(\vec{\alpha}\bullet\vec{r})^2$$
$$= r^{-2}r^2 = 1 \tag{18}$$

and,

$$\alpha_0\varepsilon + \varepsilon\alpha_0 = r^{-1}\{\alpha_0(\vec{\alpha}\bullet\vec{r})+(\vec{\alpha}\bullet\vec{r})\alpha_0\} = 0 \tag{19}$$

We shall in the first instance consider the spectral lines of hydrogen in the absence of An external magnetic field. So that (17) reduces to,

$$H = c\varepsilon p_r + ci\hbar r^{-1}\varepsilon\alpha_0 j + \alpha_0 m_0 c^2 - eV(r) \tag{20}$$

Select consistent with (18, 19),

$$\varepsilon = \begin{pmatrix} 0,-i \\ i,0 \end{pmatrix} \;,\; \alpha_0 = \begin{pmatrix} 1,0 \\ 0,-1 \end{pmatrix} \tag{21}$$

Therefore equation (20) becomes,

$$\left\{cp_r\begin{pmatrix}0,-i\\i,0\end{pmatrix}+ci\hbar r^{-1}j\begin{pmatrix}0,-i\\i,0\end{pmatrix}\begin{pmatrix}1,0\\0,-1\end{pmatrix}+m_0 c^2\begin{pmatrix}1,0\\0,-1\end{pmatrix}-eV(r)\begin{pmatrix}1,0\\0,1\end{pmatrix}\right\}\begin{pmatrix}\psi_1\\\psi_2\end{pmatrix}=E\begin{pmatrix}\psi_1\\\psi_2\end{pmatrix}$$

Using (7e),

$$(m_0c^2 - E - eV(r))\psi_1 - c\hbar\left(\frac{\partial}{\partial r} + \frac{1}{r} + \frac{j}{r}\right)\psi_2 = 0$$

(22)

$$-\hbar c\left(\frac{j}{r} - \frac{\partial}{\partial r} - \frac{1}{r}\right)\psi_1 - (m_0c^2 + E + eV(r))\psi_2 = 0$$

Try solutions of the form,

$$\psi_1(r,\theta,\phi) = \frac{f_1(r)}{r}Y(\theta,\phi) \ , \ \psi_2(r,\theta,\phi) = \frac{f_2(r)}{r}Y(\theta,\phi)$$

(23)

Substituting (23) in (22) and simplifying,

$$\left(m_0c^2 - E - \frac{e^2}{r}\right)f_1 - c\hbar\left(\frac{d}{dr} + \frac{j}{r}\right)f_2 = 0$$

(24)

$$-\hbar c\left(\frac{j}{r} - \frac{d}{dr}\right)f_1 - (m_0c^2 + E + \frac{e^2}{r})f_2 = 0$$

Define,

$$\alpha = \frac{e^2}{\hbar c}, \ c_1 = \frac{\hbar c}{m_0c^2 + E}, \ c_2 = \frac{\hbar c}{m_0c^2 - E}, \ a = \sqrt{\frac{1}{c_1c_2}} = \frac{\sqrt{m_0^2c^4 - E^2}}{\hbar c}$$

(25)

Substituting (25) in (24) gives,

$$\left(\frac{1}{c_2} - \frac{\alpha}{r}\right)f_1 - \left(\frac{d}{dr} + \frac{j}{r}\right)f_2 = 0$$

(26)

$$\left(\frac{j}{r} - \frac{d}{dr}\right)f_1 - \left(\frac{1}{c_1} + \frac{\alpha}{r}\right)f_2 = 0$$

Put,

$$f_1(r) = z_1(r)e^{-ar} \ , \ f_2(r) = z_2(r)e^{-ar}$$

(27)

Substituting (27) in (26) gives,

$$\left(\frac{1}{c_2}-\frac{\alpha}{r}\right)z_1 - \left(\frac{d}{dr}-a+\frac{j}{r}\right)z_2 = 0 \tag{28}$$

$$\left(\frac{j}{r}-\frac{d}{dr}+a\right)z_1 - \left(\frac{1}{c_1}+\frac{\alpha}{r}\right)z_2 = 0$$

Try a series expansion for $z_1(r), z_2(r)$

$$z_1(r) = \sum_{k=1}^{\infty} b_k r^{\mu+k}, \quad z_2(r) = \sum_{k=1}^{\infty} d_k r^{\mu+k} \tag{29}$$

where, $\mu+1>0$. Substituting (29) in (28) and equating to zero the coefficients of $r^{\mu+k-1}$

$$\frac{1}{c_2}b_{k-1} - \alpha b_k + ad_{k-1} - (\mu+k+j)d_k = 0 \tag{30}$$

$$ab_{k-1} - (\mu+k-j)b_k + \frac{1}{c_1}d_{k-1} + \alpha d_k = 0$$

Multiply the first of equation (30) by a and the second by $1/c_2$ and subtract and using (25),

$$\left\{\frac{\mu+k-j}{c_2} - a\alpha\right\}b_k - \left\{(\mu+k+j)a + \frac{\alpha}{c_2}\right\}d_k = 0 \tag{31}$$

From (31, 30), as $k \to \infty$,

$$\frac{b_k}{c_2} \approx ad_k \tag{32a}$$

$$\frac{1}{c_2}b_{k-1} - \alpha b_k + ad_{k-1} - kd_k \approx 0 \tag{32b}$$

From (32a, 32b),

$$ad_{k-1} - \alpha b_k + ad_{k-1} - kd_k = 0$$

$$\therefore \quad 2ad_{k-1} = kd_k + \alpha b_k = (k+\alpha c_2 a)d_k \approx kd_k \tag{32c}$$

Consider the power series,

$$e^{2ar} = \sum_{k=0}^{\infty} \frac{1}{k!}(2ar)^k = \sum_{k=0}^{\infty} u_k r^k$$

where,

$$\frac{u_{k+1}}{u_k} = \frac{(2a)^{k+1}}{(k+1)!} \frac{k!}{(2a)^k} = \frac{2a}{k+1} \approx \frac{2a}{k} \quad , \quad \text{as} \quad k \to \infty \tag{32d}$$

Therefore an examination of (32c, 32d) shows that for large k, the series (29), behaves like e^{2ar} and will consequently diverge for large r, unless (29) terminates. Therefore, there exists a σ such that the highest power in both series is, $r^{\mu+\sigma}$, with $\sigma \geq 1$. With $b_0 = d_0 = 0$, we have from (30)

$$\alpha b_1 + (\mu + 1 + j)d_1 = 0 \tag{32e}$$

$$(\mu + 1 - j)b_1 - \alpha d_1 = 0$$

Therefore for non trivial solutions,

$$-\alpha^2 - (\mu+1)^2 + j^2 = 0$$
$$\therefore \quad \mu + 1 = \sqrt{j^2 - \alpha^2} \tag{32f}$$

Again, since $b_{\sigma+1} = d_{\sigma+1}$, from (31)

$$\frac{1}{c_2} b_\sigma + a d_\sigma = 0 \tag{32g}$$

$$\left\{ \frac{\mu+\sigma-j}{c_2} - a\alpha \right\} b_\sigma - \left\{ (\mu+\sigma+j)a + \frac{\alpha}{c_2} \right\} d_\sigma = 0$$

For nontrivial solutions we have from (32g), using (25)

$$\left(\frac{(\mu+\sigma+j)a}{c_2} + \frac{\alpha}{c_2^2} \right) + \left(\frac{(\mu+\sigma-j)a}{c_2} - a^2\alpha \right) = 0$$

$$(\mu + \sigma - j - a\alpha c_2) = -\left(\mu + \sigma + j + \frac{\alpha}{ac_2}\right)$$

$$\therefore \quad 2(\mu + \sigma)a = a^2\alpha c_2 - \frac{\alpha}{c_2} = \alpha\left(\frac{1}{c_1} - \frac{1}{c_2}\right) \tag{33}$$

$$2(\mu + \sigma)\frac{\sqrt{m_0^2 c^4 - E^2}}{\hbar c} = \frac{2\varepsilon E}{\hbar c}$$

Solving for E, using (32f) we have the following expression for the energy levels of the hydrogen atom.

$$E = \frac{m_0 c^2}{\sqrt{1 + \frac{\alpha^2}{(\mu + \sigma)^2}}} = \frac{m_0 c^2}{\sqrt{1 + \frac{\alpha^2}{(\sigma - 1 + \sqrt{j^2 - \alpha^2})^2}}} \tag{34}$$

The highest power of r in the power series (29) is given by (32f),

$$N = \mu + \sigma = \sigma - 1 + \sqrt{j^2 - \alpha^2} \tag{35}$$

Define,

$$n = \sigma - 1 + |j| \quad , \quad n = \text{integer} \tag{36}$$

$\therefore$ From (34),

$$E = \frac{m_0 c^2}{\sqrt{1 + \frac{\alpha^2}{N^2}}} = m_0 c^2 \left(1 - \frac{\alpha^2}{2N^2} + \frac{3\alpha^4}{8N^4} - \ldots\right) \tag{37}$$

From (35, 36)

$$N = \sigma - 1 + |j|\left(1 - \frac{\alpha^2}{j^2}\right)^{\frac{1}{2}} = \sigma - 1 + |j|\left(1 - \frac{\alpha^2}{2j^2} - \frac{\alpha^4}{8j^4} - \ldots\right)$$

$$= \sigma - 1 + |j| - \frac{\alpha^2}{2|j|} - \frac{\alpha^4}{8|j|^3} - \ldots = n\left(1 - \frac{\alpha^2}{2n|j|} - \frac{\alpha^4}{8n|j|^3} - \ldots\right) \tag{38}$$

$$\therefore \quad \frac{1}{N^2} = \frac{1}{n^2}\left(1 + \frac{\alpha^2}{n|j|} + \frac{\alpha^4}{4n|j|} + \frac{3\alpha^4}{4n^2|j|^2} + \ldots\right) \tag{39a}$$

$$\frac{1}{N^4} = \frac{1}{n^4}\left(1 + \frac{2\alpha^2}{n|j|} + \ldots\right) \tag{39b}$$

Therefore from (37) using (39a, 39b),

$$E = m_0 c^2 \left(1 - \frac{\alpha^2}{2n^2} - \frac{\alpha^4}{2n^4}\left(\frac{n}{|j|} - \frac{3}{4}\right) - \ldots\right)$$

$$= m_0 c^2 - \frac{m_0 e^4}{\hbar^2}\left[\frac{1}{n^2} + \frac{\alpha^2}{n^4}\left(\frac{n}{|j|} - \frac{3}{4}\right) - \ldots\right] \tag{40}$$

For the transition from energy state: $E_1(n_1, j_1) \to E_2(n_2, j_2)$, the frequency of the emitted radiation is given by,

$$\therefore \; h\nu = E_2 - E_1$$

From (40),

$$\nu = \frac{m_0 e^4}{4\pi\hbar^3}\left[\left(\frac{1}{n_2^2} - \frac{1}{n_1^2}\right) + \frac{\alpha^2}{n_2^4}\left(\frac{n_2}{|j_2|} - \frac{3}{4}\right) - \frac{\alpha^2}{n_1^4}\left(\frac{n_1}{|j_1|} - \frac{3}{4}\right) - \ldots\right] \tag{41}$$

where, $|j_1|, |j_2| = 1, 2, 3, \ldots\ldots$ The first term corresponds to the Balmer lines, while the second, third,.. terms give the fine structure lines.

5. In free space in the absence of charge or current distribution, Maxwell's equations reduce to,

$$\vec{\nabla} \cdot \vec{E} = 0, \quad \vec{\nabla} \cdot \vec{H} = 0 \tag{1}$$

$$\vec{\nabla} \wedge \vec{E} = -\frac{1}{c}\frac{\partial \vec{H}}{\partial t}, \quad \vec{\nabla} \wedge \vec{H} = \frac{1}{c}\frac{\partial \vec{E}}{\partial t} \tag{2}$$

A simple mathematical manipulation gives the wave equations,

$$\nabla^2 \vec{E} = \frac{1}{c^2}\frac{\partial^2 \vec{E}}{\partial t^2}, \quad \nabla^2 \vec{H} = \frac{1}{c^2}\frac{\partial^2 \vec{H}}{\partial t^2}$$

Try solutions of the form,

$$\vec{E} = \vec{\varepsilon}^1\{a^1 e^{i(\vec{k}\bullet\vec{r}-\omega t)} + a^{*1} e^{-i(\vec{k}\bullet\vec{r}-\omega t)}\}$$

$$\vec{H} = \vec{\varepsilon}^2\{b^1 e^{i(\vec{k}\bullet\vec{r}-\omega t)} + b^{*1} e^{-i(\vec{k}\bullet\vec{r}-\omega t)}\} \tag{3}$$

where * denotes complex conjugation, and $\vec{n}$ is the direction of propagation, and

$$\vec{k} = k\vec{n} = \frac{2\pi}{\lambda}\vec{n} = \frac{2\pi\nu}{c}\vec{n} \tag{4}$$

$\vec{\varepsilon}^1, \vec{\varepsilon}^2$ in equation (3) are unit vectors which together with $\vec{n}$, constitute an orthonormal system of vectors. Equations (3) correspond to a monochromatic wave. Therefore, general solution is obtained by a superposition of these plane waves.

$$\vec{E} = \sum_{\lambda=1}^{2} \int d\vec{k}\{a_{\vec{k}}^{\lambda} e^{i(\vec{k}\bullet\vec{r}-\omega t)} + a_{\vec{k}}^{*\lambda} e^{-i(\vec{k}\bullet\vec{r}-\omega t)}\}\vec{\varepsilon}_{\vec{k}}^{\lambda}$$

$$\vec{H} = \sum_{\lambda=1}^{2} \int d\vec{k}\{b_{\vec{k}}^{\lambda} e^{i(\vec{k}\bullet\vec{r}-\omega t)} + b_{\vec{k}}^{*\lambda} e^{-i(\vec{k}\bullet\vec{r}-\omega t)}\}\vec{\varepsilon}_{\vec{k}}^{\lambda} \tag{5}$$

where, $\quad d\vec{k} = dk_x dk_y dk_z \tag{5a}$

From equations (2), using (5) we have

$$i\sum_{\lambda=1}^{2} \int d\vec{k}\{a_{\vec{k}}^{\lambda} e^{i(\vec{k}\bullet\vec{r}-\omega t)} - a_{\vec{k}}^{*\lambda} e^{-i(\vec{k}\bullet\vec{r}-\omega t)}\}(\vec{k}\wedge\vec{\varepsilon}_{\vec{k}}^{\lambda}) = \frac{i}{c}\sum_{\lambda=1}^{2} \int d\vec{k}\omega\{b_{\vec{k}}^{\lambda} e^{i(\vec{k}\bullet\vec{r}-\omega t)} - b_{\vec{k}}^{*\lambda} e^{-i(\vec{k}\bullet\vec{r}-\omega t)}\}\vec{\varepsilon}_{\vec{k}}^{\lambda}$$

$$i\sum_{\lambda=1}^{2} \int d\vec{k}\{b_{\vec{k}}^{\lambda} e^{i(\vec{k}\bullet\vec{r}-\omega t)} - b_{\vec{k}}^{*\lambda} e^{-i(\vec{k}\bullet\vec{r}-\omega t)}\}(\vec{k}\wedge\vec{\varepsilon}_{\vec{k}}^{\lambda}) = -\frac{i}{c}\sum_{\lambda=1}^{2} \int d\vec{k}\omega\{a_{\vec{k}}^{\lambda} e^{i(\vec{k}\bullet\vec{r}-\omega t)} - a_{\vec{k}}^{*\lambda} e^{-i(\vec{k}\bullet\vec{r}-\omega t)}\}\vec{\varepsilon}_{\vec{k}}^{\lambda}$$

But,
$$\vec{k}\wedge\vec{\varepsilon}_{\vec{k}}^{1} = k\vec{\varepsilon}_{\vec{k}}^{2}, \qquad \vec{k}\wedge\vec{\varepsilon}_{\vec{k}}^{2} = -k\vec{\varepsilon}_{\vec{k}}^{1} \tag{6}$$

Therefore, equating coefficients using (4),

$$a_{\vec{k}}^{2} = -b_{\vec{k}}^{1}, \quad a_{\vec{k}}^{1} = b_{\vec{k}}^{2} \tag{7}$$

From (5),

$$\vec{E} = \sum_{\lambda=1}^{2} \int d\vec{k}\{\alpha_{\vec{k}}^{\lambda} e^{i\vec{k}\bullet\vec{r}} + \alpha_{\vec{k}}^{*\lambda} e^{-i\vec{k}\bullet\vec{r}}\}\vec{\varepsilon}_{\vec{k}}^{\lambda}$$

$$\vec{H} = \sum_{\lambda=1}^{2} \int d\vec{k}\{\beta_{\vec{k}}^{\lambda} e^{i\vec{k}\bullet r} + \beta_{\vec{k}}^{*\lambda} e^{-i\vec{k}\bullet\vec{r}}\}\vec{\varepsilon}_{\vec{k}}^{\lambda} \tag{8}$$

where,

$$\alpha_{\vec{k}}^{\lambda} = a_{\vec{k}}^{\lambda} e^{-i\omega t} \quad , \quad \beta_{\vec{k}}^{\lambda} = b_{\vec{k}}^{\lambda} e^{-i\omega t} \tag{9}$$

Therefore, from (8)

$$E^2 = \sum_{\lambda}\sum_{\lambda'} \iint d\vec{k}d\vec{k}' \vec{\varepsilon}_{\vec{k}}^{\lambda} \bullet \vec{\varepsilon}_{\vec{k}'}^{\lambda'} \{\alpha_{\vec{k}}^{\lambda}\alpha_{\vec{k}'}^{\lambda'} e^{i(\vec{k}+\vec{k}')\bullet\vec{r}} + \alpha_{\vec{k}}^{*\lambda}\alpha_{\vec{k}'}^{\lambda'} e^{-i(\vec{k}-\vec{k}')\bullet\vec{r}}$$

$$+ \alpha_{\vec{k}}^{\lambda}\alpha_{\vec{k}'}^{*\lambda'} e^{i(\vec{k}-\vec{k}')\bullet\vec{r}} + \alpha_{\vec{k}}^{*\lambda}\alpha_{\vec{k}'}^{*\lambda'} e^{-i(\vec{k}+\vec{k}')\bullet\vec{r}}\} \tag{10}$$

$$H^2 = \sum_{\lambda}\sum_{\lambda'} \iint d\vec{k}d\vec{k}' \vec{\varepsilon}_{\vec{k}}^{\lambda} \bullet \vec{\varepsilon}_{\vec{k}'}^{\lambda'} \{\beta_{\vec{k}}^{\lambda}\beta_{\vec{k}'}^{\lambda'} e^{i(\vec{k}+\vec{k}')\bullet\vec{r}} + \beta_{\vec{k}}^{*\lambda}\beta_{\vec{k}'}^{\lambda'} e^{-i(\vec{k}-\vec{k}')\bullet\vec{r}} + \beta_{\vec{k}}^{\lambda}\beta_{\vec{k}'}^{*\lambda'} e^{i(\vec{k}-\vec{k}')\bullet\vec{r}}$$

$$+ \beta_{\vec{k}}^{*\lambda}\beta_{\vec{k}'}^{*\lambda'} e^{-i(\vec{k}+\vec{k}')\bullet\vec{r}}\}$$

Using the result,

$$\frac{1}{2\pi}\int_{-\infty}^{\infty} e^{-i\sigma T} d\sigma = \delta(T) \tag{11}$$

We have from (10),

$$\int E^2 d\vec{r} = 8\pi^3$$

$$\sum_{\lambda}\sum_{\lambda'} \iint d\vec{k}d\vec{k}' \vec{\varepsilon}_{\vec{k}}^{\lambda} \bullet \vec{\varepsilon}_{\vec{k}'}^{\lambda'} \{(\alpha_{\vec{k}}^{\lambda}\alpha_{\vec{k}'}^{\lambda'} + \alpha_{\vec{k}}^{*\lambda}\alpha_{\vec{k}'}^{*\lambda'})\delta(\vec{k}+\vec{k}')$$

$$+ (\alpha_{\vec{k}}^{*\lambda}\alpha_{\vec{k}'}^{\lambda'} + \alpha_{\vec{k}}^{\lambda}\alpha_{\vec{k}'}^{*\lambda'})\delta(\vec{k}-\vec{k}')\} \tag{12}$$

$$\int H^2 d\vec{r} = 8\pi^3 \sum_{\lambda}\sum_{\lambda'}\iint d\vec{k}d\vec{k}'\vec{\varepsilon}_{\vec{k}}^{\lambda}\bullet\vec{\varepsilon}_{\vec{k}'}^{\lambda'}\{(\beta_{\vec{k}}^{\lambda}\beta_{\vec{k}'}^{\lambda'}+\beta^{*\lambda}_{\vec{k}}\beta^{*\lambda'}_{\vec{k}'})\delta(\vec{k}+\vec{k}')$$
$$+(\beta^{*\lambda}_{\vec{k}}\beta_{\vec{k}'}^{\lambda'}+\beta_{\vec{k}}^{\lambda}\beta^{*\lambda'}_{\vec{k}'})\delta(\vec{k}-\vec{k}')\}$$

where,

$$dV = d\vec{r} = dxdydz$$

Using the result,

$$\int_{-\infty}^{\infty} f(x)\delta(x-a) = f(a) \tag{13}$$

We have from (12),

$$\int E^2 d\vec{r} = 8\pi^3 \sum_{\lambda=1}^{2}\sum_{\lambda'=1}^{2}\int d\vec{k}\{\vec{\varepsilon}_{\vec{k}}^{\lambda}\bullet\vec{\varepsilon}_{-\vec{k}}^{\lambda'}(\alpha_{\vec{k}}^{\lambda}\alpha_{-\vec{k}}^{\lambda'}+\alpha^{*\lambda}_{\vec{k}}\alpha^{*\lambda'}_{-\vec{k}})+\vec{\varepsilon}_{\vec{k}}^{\lambda}\bullet\vec{\varepsilon}_{\vec{k}}^{\lambda'}(\alpha^{*\lambda}_{\vec{k}}\alpha_{\vec{k}}^{\lambda'}+\alpha_{\vec{k}}^{\lambda}\alpha^{*\lambda'}_{\vec{k}})\}$$

Defining,

$$\vec{\varepsilon}_{-\vec{k}}^{1} = \vec{\varepsilon}_{\vec{k}}^{1}, \quad \vec{\varepsilon}_{-\vec{k}}^{2} = -\vec{\varepsilon}_{\vec{k}}^{2} \tag{14}$$

and using the condition,

$$\vec{\varepsilon}_{\vec{k}}^{\lambda}\bullet\vec{\varepsilon}_{\vec{k}}^{\lambda'} = \delta_{\lambda\lambda'} \tag{15}$$

$$\int E^2 d\vec{r} = 8\pi^3 \sum_{\lambda=1}^{2}\sum_{\lambda'=1}^{2}\int d\vec{k}\{\vec{\varepsilon}_{\vec{k}}^{\lambda}\bullet\vec{\varepsilon}_{-\vec{k}}^{\lambda'}(\alpha_{\vec{k}}^{\lambda}\alpha_{-\vec{k}}^{\lambda'}+\alpha^{*\lambda}_{\vec{k}}\alpha^{*\lambda'}_{-\vec{k}})+\vec{\varepsilon}_{\vec{k}}^{\lambda}\bullet\vec{\varepsilon}_{\vec{k}}^{\lambda'}(\alpha^{*\lambda}_{\vec{k}}\alpha_{\vec{k}}^{\lambda'}+\alpha_{\vec{k}}^{\lambda}\alpha^{*\lambda'}_{\vec{k}})\}$$

$$= 8\pi^3 \int d\vec{k}\{(a_{\vec{k}}^{1}a_{-\vec{k}}^{1}-a_{\vec{k}}^{2}a_{-\vec{k}}^{2})e^{-2i\omega t}+(a^{*1}_{\vec{k}}a^{*1}_{-\vec{k}}-a^{*2}_{\vec{k}}a^{*2}_{-\vec{k}})e^{2i\omega t}\}+16\pi^3 \sum_{\lambda}\int d\vec{k}(a^{*\lambda}_{\vec{k}}a_{\vec{k}}^{\lambda})$$

Since the oscillatory terms vanish when averaged over time, we have

$$<\int E^2 d\vec{r}> = 16\pi^3 \sum_{\lambda}\int d\vec{k}(a^{*\lambda}_{\vec{k}}a_{\vec{k}}^{\lambda}) \tag{16}$$

Similarly,

$$<\int H^2 d\vec{r}> = 16\pi^3 \sum_{\lambda}\int d\vec{k}(b^{*\lambda}_{\vec{k}}b_{\vec{k}}^{\lambda}) \tag{17}$$

Therefore, the mean energy of the electromagnetic field is using

$$W = <\frac{1}{8\pi}\int(E^2+H^2)d\vec{r}> = 4\pi^2\sum_{\lambda}(a^{*\lambda}_{\vec{k}}a^{\lambda}_{\vec{k}}) \qquad (18)$$

In the transition to quantum theory, the c- numbers ($a^{\lambda}_{\vec{k}}, a^{*\lambda}_{\vec{k}}$) are represented by linear operators, ($a^{\lambda}_{\vec{k}}, a^{+\lambda}_{\vec{k}}$).

For a system of charged particles $\{q_s\}$ of rest masses $\{m_{0s}\}$, the total Hamiltonian is given by,

$$H = W + \sum_s c\sqrt{m_{0s}^2 c^2 + |\vec{p}_s - \frac{q_s}{c}\vec{A}_s|^2} + \frac{1}{2}\sum_s\sum_{s\neq m}\frac{q_s q_m}{|\vec{r}_s - \vec{r}_m|} \qquad (19)$$

Using Dirac's theory, for a system of electrons, the middle term in equation (19) can be written as,

$$H_{el} = \sum_s \{c[\vec{\alpha}_s \bullet (\vec{p}_s + \frac{e}{c}\vec{A}_s) + \alpha_{0s}m_{0s}c^2]\} \qquad (20)$$

6. Integrating by parts,

$$\int_{-\infty}^{\infty}\frac{\sin^2 x}{x^2}dx = \left[-\frac{\sin^2 x}{x}\right]_{-\infty}^{+\infty} + \int_{-\infty}^{+\infty}2\frac{\sin x \cos x}{x}dx = \int_{-\infty}^{+\infty}\frac{\sin 2x}{x}dx = \int_{-\infty}^{+\infty}\frac{\sin x}{x}dx \qquad (21)$$

Let,

$$S = \int_{-\infty}^{+\infty}\frac{\sin x}{x}dx$$

$$C = \int_{-\infty}^{+\infty}\frac{\cos x}{x}dx$$

$$\therefore \quad C+iS = \int_{-\infty}^{+\infty}\frac{\cos x + i\sin x}{x}dx = \int_{-\infty}^{+\infty}\frac{e^{ix}}{x}dx$$

Consider the contour integral,

$$I = \oint_C \frac{e^{iz}}{z}dz \qquad , z = x+iy \quad , z = re^{i\theta} \qquad (22)$$

where C is the closed contour,

$$C = C_x \cup C_\rho \cup C_R$$

C_x = real axis between $x = -R \to -\rho$ and $x = \rho \to R$

C_ρ = semi circle of radius ρ, and $\theta = \pi \to 0$, $z = \rho e^{i\theta}$

C_R = semi circle of radius R and $\theta = 0 \to \pi$, $z = Re^{i\theta}$

where, $R \to \infty, \rho \to 0$

$\therefore$ From (22),

$$I = [(\int_{-R}^{-\rho} \frac{e^{ix}}{x} dx + \int_{\rho}^{R} \frac{e^{ix}}{x} dx) + \int_{\pi}^{0} \frac{e^{i\rho(\cos\theta + i\sin\theta)}}{\rho e^{i\theta}}(\rho i e^{i\theta} d\theta) + \int_{0}^{\pi} \frac{e^{iR(\cos\theta + i\sin\theta)}}{Re^{i\theta}}(Rie^{i\theta} d\theta)]$$

$$= [(\int_{-R}^{-\rho} \frac{e^{ix}}{x} dx + \int_{\rho}^{R} \frac{e^{ix}}{x} dx) + \int_{\pi}^{0} ie^{i\rho(\cos\theta + i\sin\theta)} d\theta + \int_{0}^{\pi} ie^{iR(\cos\theta + i\sin\theta)} d\theta] \quad (23)$$

As, $R \to \infty, \rho \to 0$

$e^{i\rho(\cos\theta + i\sin\theta)} \to 1$

$|e^{iR(\cos\theta + i\sin\theta)}| \equiv e^{-R\sin\theta} \to 0$, where $\theta = 0 \to \pi$

Therefore from (23),

$$I = \int_{-\infty}^{+\infty} \frac{e^{ix}}{x} dx - i\pi \equiv 0 \text{, since there are no poles within } C$$

Therefore,

$$\int_{-\infty}^{+\infty} \frac{e^{ix}}{x} dx = i\pi \quad (24)$$

Equating real and imaginary parts in (24),

$$\int_{-\infty}^{+\infty} \frac{\cos x}{x} dx = 0 \quad (25)$$

$$\int_{-\infty}^{+\infty} \frac{\sin x}{x} dx = \pi \tag{26}$$

7. Let, H_0 = Hamiltonian for unperturbed system

   H' = Hamiltonian for perturbed system

   H = Hamiltonian for the composite system

   $\therefore \quad H = H_0 + H' \tag{27}$

Let, $\{\psi_1, \psi_2, \psi_3, ..., \psi_n, ...\}$ be the time independent normalized wave functions of H_0.

$$H_0 \psi_n = E_n \psi_n \tag{28}$$

$$\int \psi_m^* \psi_n = \delta_{mn} \tag{28a}$$

where the integral sign denotes both an integration over continuous variables and a summation over a discrete spectrum of values.

In addition,

$$i\hbar \frac{\partial \Psi}{\partial t} = H_0 \Psi \tag{28b}$$

Try solutions of the form,

$$\Psi = \sum_n b_n \psi_n e^{-\frac{iE_n t}{\hbar}} \tag{29}$$

where, b_n = constants, normalized such that

$$\sum_n |b_n|^2 = 1 \tag{30}$$

Suppose the interaction is switched on at time $t = 0$. The wave equation now becomes,

$$i\hbar \frac{\partial \Phi}{\partial t} = (H_0 + H')\Phi \tag{31}$$

Try solutions of the form,

$$\Phi = \sum_n b_n(t)\psi_n e^{-i\frac{E_n}{\hbar}t} \qquad (32)$$

where,

$$b_n(0) = b_n \qquad (32a)$$

Substituting (32) in (31) and simplifying using the orthonormality condition (28a) gives,

$$i\hbar \dot{b}_m(t) = \sum_n H'_{mn} b_n(t) e^{i\frac{(E_m-E_n)t}{\hbar}} \qquad (33)$$

where the matrix elements of the interaction Hamiltonian is defined by,

$$H'_{mn} = \int \psi_m^* H' \psi_n \qquad (34)$$

Since H' is self-adjoint we have,

$$H'_{mn} = H'^{*}_{nm} \qquad (35)$$

Again from (33),

$$i\hbar \frac{d}{dt}\sum_m |b_m|^2 = i\hbar \sum_m (\dot{b}_m b_m^* + b_m \dot{b}_m^*)$$

$$= \sum_m \sum_n H'_{mn} b_n(t) b_m^*(t) e^{i\frac{(E_m-E_n)t}{\hbar}} - \sum_m \sum_n b_m(t) H'^{*}_{mn} b_n^*(t) e^{-i\frac{(E_m-E_n)}{\hbar}}$$

$$= \sum_m \sum_n H'_{mn} b_n(t) b_m^*(t) e^{i\frac{(E_m-E_n)t}{\hbar}} - \sum_n \sum_m b_n(t) H'^{*}_{nm} b_m^*(t) e^{-i\frac{(E_n-E_m)t}{\hbar}}$$

interchanging the dummy suffices (m, n) in the second summation. From (35), it therefore follows,

$$i\hbar \frac{d}{dt}\sum_m |b_m(t)|^2 = 0 \qquad (36)$$

Therefore,

$$\sum_m |b_m(t)|^2 = \sum_m |b_m(0)|^2 = 1 \qquad (37)$$

This shows that the $b_m(t)$ are normalized, and the probability that the system is in state m is $|b_m(t)|^2$. To solve the system of equations (33) we make the following assumptions.

- The interaction is switched on at time $t = 0$, and switched off a short time t later. The system is initially in a definite state labeled '0". So that,

$$b_n(0) = \delta_{n0} \qquad (38)$$

- For t sufficiently small,

$$b_n(t) \approx b_n(0) \qquad (38a)$$

Substituting (38, 38a) in (33),

$$i\hbar \dot{b}_m(t) \approx \sum_n H'_{mn} b_n(0) e^{i\frac{(E_m-E_n)t}{\hbar}} = H'_{m0} e^{i\frac{(E_m-E_0)t}{\hbar}} \qquad (39)$$

Integrating (39),

$$b_m(t) = \frac{H'_{m0}}{E_m - E_0}\left[1 - e^{i\frac{(E_m-E_0)t}{\hbar}}\right], \quad m \neq 0 \qquad (40)$$

When $H'_{m0} = 0$ a direct transition between states $((0 \to m)$ is not possible. Instead one has to transit via an intermediate state l, for which $H'_{l0}, H'_{ml} \neq 0$.

Virtual States :

$0 \to l \to m$; intermediate state (l) = virtual state, between transition from an initial state (0) to a final state (m)

The procedure is as follows: From (39),

$$i\hbar \dot{b}_l(t) = H'_{l0} e^{i\frac{(E_l-E_0)t}{\hbar}} \qquad (41a)$$

Integrating (41),

$$b_l(t) = -\frac{H'_{l0}}{E_l - E_0} e^{i\frac{(E_l-E_0)t}{\hbar}} \quad , \quad b_l(0) \neq 0 \tag{41b}$$

Substitute (41b) in (33),

$$i\hbar \dot{b}_m(t) = \sum_l H'_{ml} b_l(t) e^{i\frac{(E_m-E_l)t}{\hbar}} = \sum_l \frac{H'_{ml} H'_{l0}}{E_0 - E_l} e^{i\frac{(E_m-E_0)t}{\hbar}} \tag{41c}$$

Integrating (41c),

$$b_m(t) = \sum_l \frac{H'_{ml} H'_{l0}}{E_0 - E_l} \frac{1 - e^{i\frac{(E_m-E_0)t}{\hbar}}}{E_m - E_0} \quad ; \quad b_m(0) = 0 \tag{41d}$$

If either of the above interaction matrix elements vanish we may be forced to go thru' 3 or more virtual states: For example,

$$0 \to l \to r \to m$$

where (l, r) are virtual states. We again proceed as follows:

From (41c),

$$i\hbar \dot{b}_r(t) = \sum_l \frac{H'_{rl} H'_{l0}}{E_0 - E_l} e^{i\frac{(E_r-E_0)t}{\hbar}} \tag{42a}$$

Integrating,

$$b_r(t) = \sum_l \frac{H'_{rl} H'_{l0}}{E_0 - E_l} \frac{e^{i\frac{(E_r-E_0)t}{\hbar}}}{E_0 - E_r} \quad , \quad b_r(0) \neq 0 \tag{42b}$$

Substituting (42b) in (33)

$$i\hbar \dot{b}_m(t) = \sum_r H'_{mr} b_r(t) e^{i\frac{(E_m-E_r)t}{\hbar}} = \sum_r \sum_l \frac{H'_{mr} H'_{rl} H'_{l0}}{(E_0 - E_l)(E_0 - E_r)} e^{i\frac{(E_m-E_0)t}{\hbar}} \tag{42c}$$

Integrating,

$$b_m(t) = \sum_r \sum_l \frac{H'_{mr} H'_{rl} H'_{l0}}{(E_0 - E_l)(E_0 - E_r)} \frac{1 - e^{i\frac{(E_m-E_0)t}{\hbar}}}{(E_m - E_0)} \quad , \quad b_m(0) = 0 \tag{42d}$$

From (40, 41d, 42d) we can write,

$$b_m(t) = H' \frac{1 - e^{i\frac{(E_m - E_0)t}{\hbar}}}{E_m - E_0} \qquad (43)$$

where,

first order: $\qquad H' = H'_{m0} \qquad (44a)$

second order: $\qquad H' = \sum_l \frac{H'_{ml} H'_{l0}}{E_0 - E_l} \qquad (44b)$

third order: $\qquad H' = \sum_r \sum_l \frac{H'_{mr} H'_{rl} H'_{l0}}{(E_0 - E_l)(E_0 - E_r)} \qquad (44c)$

From (43), the probability that the system will be in the state m at time t is

$$|b_m(t)|^2 = |H'|^2 \left| \frac{e^{-i\frac{(E_m - E_0)t}{2\hbar}} - e^{+i\frac{(E_m - E_0)t}{2\hbar}}}{E_m - E_0} \right|^2 = |H'|^2 \frac{\sin^2\left(\frac{E_m - E_0}{2\hbar} t\right)}{(E_m - E_0)^2} \qquad (45)$$

The number of states whose energy lies between $(E_m, E_m + dE_m) = \rho_{E_m} dE_m \qquad (46)$
Therefore the probability the system is in one of these states at time t is,

$$\int dE_m \, \rho_{E_m} |H'|^2 \frac{\sin^2\left(\frac{E_m - E_0}{2\hbar} t\right)}{(E_m - E_0)^2} = \frac{2t}{\hbar} \int \rho_{E_m} |H'|^2 \frac{\sin^2 x}{x^2} dx \, , \quad x = \frac{E_m - E_0}{2\hbar} t$$

Using the result from exercise 6 we can approximate, the above integral by,
$\frac{2t\pi}{\hbar} |H'|^2 \rho_F$, where ρ_F is the density function for the final state. Therefore,
The transition probability per unit time is given by,

$$w = \frac{2\pi}{\hbar} |H'|^2 \rho_F \qquad (47)$$

8. Dirac's Assumptions :

   (1) The fundamental building blocks of the physical universe: Planck's constant (h), velocity of light in a vacuum (c), Newton's gravitational constant (G), Hubble's parameter (H).

   (2) H, G depend on time, but their ratio is independent of time.

   (3) Particles whose rest mass is time independent constitute stable particles.

The ratio of the electrostatic to the gravitational forces between an electron and a proton is,

$$\frac{e^2/r^2}{Gm_e m_p/r^2} \approx 10^{40} \tag{45}$$

The present age of the universe is given by the reciprocal of the present value of Hubble's parameter.

$$t_0 = \frac{1}{H_0} \approx 10^{17} \text{ sec} \tag{45a}$$

The classical electron radius is : $r_0 = \dfrac{e^2}{m_e c^2}$ \hfill (45b)

The atomic unit of time is : $t_{at} = \dfrac{r_0}{c} = \dfrac{e^2}{m_e c^3} \approx 10^{-23}$ \hfill (45c)

∴ from (45a, 45c)

$$\frac{t_0}{t_{at}} \approx 10^{-40} \tag{45d}$$

Therefore, from (45, 45d)

$$\frac{e^2}{Gm_e m_p} \approx \frac{t_0}{t_{at}} \Rightarrow G = \frac{e^4}{t_0 m_e m_p c^3} \tag{45e}$$

Using (45a),

$$G(t_0) = \frac{e^4}{m_e m_p c^3} H(t_0) \qquad \Rightarrow G(t_0) \propto H(t_0) \tag{45f}$$

Using dimensional analysis, particle mass can be expressed in terms of the fundamental parameters (1) that describe the universe,

$$[M] = h^\alpha H^\beta G^\gamma c^\delta = [ML^2T^{-1}]^\alpha [T^{-1}]^\beta [M^{-1}L^3T^{-2}]^\gamma [LT^{-1}]^\delta$$

$$= M^{\alpha-\gamma} L^{2\alpha+3\gamma+\delta} T^{-\alpha-\beta-2\gamma-\delta} \tag{46}$$

Equating corresponding exponents,

$$\alpha - \gamma = 1$$

$$2\alpha + 3\gamma + \delta = 0 \tag{46a}$$

$$-\alpha - \beta - 2\gamma - \delta = 0$$

Solving,

$$\alpha = \frac{3-\delta}{5}, \quad \beta = \frac{1-2\delta}{5}, \quad \gamma = -\frac{2+\delta}{5} \tag{46b}$$

Therefore, the mass of an elementary particle can be written as,

$$m^5 = h^{3-\delta} H^{1-2\delta} G^{-2-\delta} c^{5\delta} \tag{47}$$

Equation (47) can be written in the equivalent form,

$$m^5 = \frac{h^3 H}{G^2} \left(\frac{c^5}{hH^2 G} \right)^\delta$$

$$\therefore \qquad m(b) = k(b) \left(\frac{h^3 H}{G^2} \right)^{\frac{1}{5}} \left(\frac{c^5}{hH^2 G} \right)^{\frac{b}{15}} \tag{47a}$$

where,

$$b = 3\delta \tag{47b}$$

and *k(b)* is a dimensionless constant. From (45f),

$$m(b) \propto G^{-\frac{b+1}{5}} \tag{47c}$$

Therefore for stable particles,

$$b = -1 \tag{47d}$$

9. (a) Particle model of the universe:

galaxies of masses $(m_1, m_2, m_3, ...)$ situated at time t, at position vectors $\{\vec{r}_1(t), \vec{r}_2(t), \vec{r}_3(t), ...\}$ respectively with respect to a fixed origin O. The galaxies are moving in radial directions as observed from O.

Then the total energy is,

$$E = \frac{1}{2}\sum_{i=1}^{n} m_i \dot{r}_i^2 - G\sum_{i=1}^{n}\sum_{(i<j)j=1}^{n} \frac{m_i m_j}{|\vec{r}_i - \vec{r}_j|} - \frac{\lambda}{6}\sum_{i=1}^{n} m_i r_i^2 \tag{48}$$

which includes a cosmological term, corresponding to a radial force,

$$F_{ci} = \frac{\lambda}{3} m_i r_i \tag{48a}$$

But, from Hubble's Law,

$$\vec{r}_i(t) = R(t)\vec{r}_i(t_0) \quad , \quad R(t) = \exp(\int_{t_0}^{t} H(t)dt) \tag{48b}$$

From (48b),

$$\dot{\vec{r}}_i(t) = \dot{R}(t)\vec{r}_i(t_0) = R(t)H(t)\vec{r}_i(t_0) = H(t)\vec{r}_i(t) \tag{48c}$$

and,

$$H(t) = \frac{\dot{R}(t)}{R(t)} \tag{48d}$$

Substituting (48b, 48c, 48d) in (48)

$$E = \frac{1}{2}\sum_{i=1}^{n} m_i r_i^2(t_0) \dot{R}^2(t) - G\sum_{i=1}^{n}\sum_{(i<j)j=1}^{n} \frac{m_i m_j}{R(t)|\vec{r}_i(t_0) - \vec{r}_j(t_0)|} - \frac{\lambda}{6}\sum_{i=1}^{n} m_i r_i^2(t_0) R^2(t)$$

$$= A\dot{R}^2(t) - \frac{B}{R(t)} - DR^2(t) \tag{48e}$$

where,

$$A = \frac{1}{2}\sum_{i=1}^{n} m_i r_i^2(t_0)$$

$$B = G\sum_{i=1}^{n}\sum_{(i<j)j=1}^{n} \frac{m_i m_j}{|\vec{r}_i(t_0) - \vec{r}_j(t_0)|}$$

(48f)

$$D = \frac{\lambda}{6}\sum_{i=1}^{n} m_i r_i^2(t_0)$$

From (48f),

$$\frac{D}{A} = \frac{\lambda}{3}$$

(48g)

Using (48g) in (48e),

$$\dot{R}^2(t) = \frac{B}{AR(t)} + \frac{\lambda}{3}R^2(t) + \frac{E}{A}$$

(48h)

Since our main interest is in studying the evolution of the universe, let us rescale the differential equation as follows,

$$\Re(t) = PR(t)$$

(48i)

where P is independent of time. From (48b),

$$\vec{r}_i(t) = \frac{\Re(t)}{P}\vec{r}_i(t_0) = \vec{s}_i \Re(t) \quad , \quad \vec{s}_i(t_0) = \frac{\vec{r}_i(t_0)}{P}$$

(48j)

where $\vec{s}_i$ is a constant in time for each galaxy. Substituting (48i) in (48h),

$$\frac{\dot{\Re}^2(t)}{P^2} = \frac{BP}{A}\frac{1}{\Re(t)} + \frac{\lambda}{3P^2}\Re^2(t) + \frac{E}{A}$$

$$\therefore \quad \dot{\Re}^2(t) = \frac{BP^3}{A}\frac{1}{\Re(t)} + \frac{\lambda}{3}\Re^2(t) + \frac{EP^2}{A}$$

(48k)

To obtain agreement with general relativity, we choose P such that,

(i) $E > 0$, choose $P^2 = \dfrac{c^2 A}{E} \Rightarrow \dfrac{EP^2}{A} = -kc^2, k = -1$

(ii) $E < 0$, choose $P^2 = \dfrac{c^2 A}{|E|} \Rightarrow \dfrac{EP^2}{A} = -kc^2, k = 1$

(iii) $E = 0$, choose P arbitrary, $\dfrac{EP^2}{A} = -kc^2, k = 0$

This implies,

$$P^2 = \dfrac{c^2 A}{|E|}, E \neq 0 \tag{48l}$$

The cosmological differential equation (48k) becomes,

$$\dot{\mathfrak{R}}^2(t) = \dfrac{C}{\mathfrak{R}(t)} + \dfrac{\lambda}{3}\mathfrak{R}^2(t) - kc^2 \tag{48m}$$

where from (48l),

$$C = \dfrac{BP^3}{A} = \dfrac{BPc^2}{|E|} \tag{48n}$$

The rate at which expansion of the universe slows down is the *deceleration parameter* defined by,

$$q = -\dfrac{R(t)\ddot{R}(t)}{\dot{R}^2(t)} = -\dfrac{\mathfrak{R}(t)\ddot{\mathfrak{R}}(t)}{\dot{\mathfrak{R}}^2(t)} \tag{48o}$$

(b) continuum model : The universe is represented by a sphere center O, and whose radius at time t is $a(t)$. Then, at initial time t_0, equations (48f) goes over into, $\sum \to \int$

$$A = \dfrac{1}{2}\sum_{i=1}^{n} m_i r_i^2(t_0) \Rightarrow \dfrac{1}{2}\int_0^{a(t_0)} 4\pi x^2 x^2 dx \rho_0 = \dfrac{2\pi \rho_0}{5} a_0^5 = \dfrac{3}{10} M_T a_0^2 \tag{49}$$

$$B = G\sum_{i=1}^{n} \sum_{(i<j)j=1}^{n} \dfrac{m_i m_j}{|\vec{r}_i(t_0) - \vec{r}_j(t_0)|}$$

$$= -\sum_{i=1}^{n} m_i \Phi(\vec{r}_i(t_0)) = -\int_0^{a_0} 4\pi x^2 dx \rho_0 \left[-\frac{GM(x)}{x} \right] = 4\pi \rho_0 G \int_0^{a_0} x^2 \frac{\frac{4}{3}\pi x^3 \rho_0}{x} dx$$

$$= \frac{16\pi^2 \rho_0^2 G a_0^5}{15} \tag{49a}$$

where,

$\rho_0 = \rho(t_0)$ → matter density at time $t = t_0$

$a_0 = a(t_0)$ → radius of the universe at initial time $t = t_0$

$$M_T = \int_0^{a_0} 4\pi x^2 dx \rho_0 = \frac{4\pi a_0^3}{3} \rho_0 \rightarrow \text{total matter in the spherical universe}$$

of radius a_0 at $t = t_0$

$$M(x) = \int_0^x 4\pi y^2 dy \rho_0 = \text{mass within a sphere of radius, } x$$

$\Phi(\vec{r}_i(t_0)) = $ gravitational potential at $\vec{r}_i(t_0)$

From (48n, 49, 49a),

$$C = \frac{8\pi G \rho_0 P^3}{3} \tag{49b}$$

Again from (48b),

$$a(t) = R(t)a(t_0) = R(t)a_0$$

Since the mass within the spherical universe is conserved,

$$\frac{\rho_0}{\rho} = \left(\frac{a}{a_0}\right)^3 = R^3(t)$$

Therefore (49b) yields using (48i),

$$C = \frac{8\pi G \rho \mathfrak{R}^3(t)}{3} \tag{49c}$$

Therefore for the case of a continuous distribution of matter in the universe, the cosmological differential equation is still,

$$\dot{\mathfrak{R}}^2(t) = \frac{C}{\mathfrak{R}(t)} + \frac{\lambda}{3}\mathfrak{R}^2(t) - kc^2 \qquad (48m)$$

with C given by (49c). From (48d, 48i),

$$H = \frac{\dot{\mathfrak{R}}}{\mathfrak{R}} \qquad (49d)$$

differentiating (48m) gives,

$$\ddot{\mathfrak{R}} = -\frac{C}{2\mathfrak{R}^2} + \frac{\lambda}{3\mathfrak{R}} \qquad (49e)$$

10. (a) $\lambda = 0$, from (48m, 49c),

$$kc^2 = -\dot{\mathfrak{R}}^2(t) + \frac{C}{\mathfrak{R}(t)} = -H^2(t)\mathfrak{R}^2(t) + \frac{8\pi G\rho}{3}\mathfrak{R}^2(t) \qquad (50)$$

$$= \frac{8\pi G \mathfrak{R}^2}{3}(\rho - \rho_c) \;,\; \text{where} \;\; \rho_c = \frac{3H^2}{8\pi G} \qquad (50a)$$

If at some cosmological time t, $\rho > \rho_c$, $k = 1$ for all time. $\mathfrak{R}$ reaches a maximum given by (49e),

$$\dot{\mathfrak{R}} = 0, \ddot{\mathfrak{R}} < 0 \;\Rightarrow\; \mathfrak{R} \equiv \mathfrak{R}_{max} = \frac{C}{c^2} \qquad (50b)$$

this is followed by a contraction. In other words we have an oscillating universe. On the other hand, if at some cosmological time t, $\rho < \rho_c$, $k = -1$ for all time, and from (50), $\dot{\mathfrak{R}}^2 > 0$ → permanent expansion. Again if $\rho = \rho_c$, $k = 0$, $\dot{\mathfrak{R}}^2 > 0$ → permanent expansion

(b) Einstein- de Sitter universe : $k = \lambda = 0$
   Equation (50) becomes,

$$0 = -\dot{\mathfrak{R}}^2(t) + \frac{C}{\mathfrak{R}(t)} \;\Rightarrow\; \mathfrak{R}^{\frac{1}{2}} d\mathfrak{R} = C^{\frac{1}{2}} dt$$

Therefore, integrating and simplifying

$$\mathfrak{R}^3 = \frac{9}{4}Ct^2 \;,\; \text{with} \;\; \mathfrak{R} = 0, t = 0 \;\Rightarrow\; \text{at the moment of the big-bang the radius of the universe was zero.} \qquad (51)$$

Substituting from (49c),

$$\rho = \frac{1}{6\pi G t^2} \tag{51a}$$

From (50a, 51a),

$$\rho = \rho_c = \frac{3H^2}{8\pi G} \quad \Rightarrow \quad H = \frac{2}{3t} \tag{51b}$$

From (48o), in conjunction with (48d, 49e),

$$q = -\frac{1}{\Re H^2}\left(-\frac{C}{2\Re^2} + \frac{\lambda}{3}\Re\right) \tag{51c}$$

For the Einstein- de Sitter model, this reduces to using (51),

$$q = \frac{C}{2H^2 \Re^3} = \frac{1}{2} \tag{51d}$$

References:

P. T. Landsberg & D. A. Evans – Mathematical Cosmology, An Introduction, Clarendon Press, Oxford, 1977; Chapters 4-6, Appendix B.

J. McConnell – Quantum Particle Dynamics, North-Holland Publishing Co., 1960; Chapter 8

J. R. Oppenheimer – Lectures on Electrodynamics; Gordon & Breach, 1970

www.ingramcontent.com/pod-product-compliance
Lightning Source LLC
Chambersburg PA
CBHW081112170526
45165CB00008B/2426